Compounds with Polar Metallic Bonding

Compounds with Polar Metallic Bonding

Special Issue Editor

Constantin Hoch

MDPI • Basel • Beijing • Wuhan • Barcelona • Belgrade

MDPI

Special Issue Editor
Constantin Hoch
LMU Munich, Department for Chemistry,
Germany

Editorial Office
MDPI
St. Alban-Anlage 66
4052 Basel, Switzerland

This is a reprint of articles from the Special Issue published online in the open access journal *Crystals* (ISSN 2073-4352) from 2018 to 2019 (available at: https://www.mdpi.com/journal/crystals/special_issues/Polar_Metallic_Bonding)

For citation purposes, cite each article independently as indicated on the article page online and as indicated below:

LastName, A.A.; LastName, B.B.; LastName, C.C. Article Title. *Journal Name* **Year**, *Article Number*, Page Range.

ISBN 978-3-03921-070-1 (Pbk)
ISBN 978-3-03921-071-8 (PDF)

Contents

About the Special Issue Editor

Constantin Hoch studied Chemistry at the University of Freiburg im Breisgau (Germany), where he also received his PhD degree in 2003. First as a postdoc, then as scientific coworker, he stayed for several years at the Max Planck Institute for Solid State Research in Stuttgart (Germany). After a short stay at Stuttgart University, he moved to Munich University, where he earned his Habilitation in 2018. His research interests include transition forms of metallic and ionic bonding and compound classes such as subvalent alkaline and earth alkaline metal compounds, amalgams of less noble metals and metal-rich metalates. These chemical systems require development of modern preparation methods, X-ray crystallography on highly absorbing materials, and DFT calculations of the electronic structures of solids with polar metal-metal bonding. His results have been published in internationally renowned peer-reviewed journals (105 publications to date, with 677 citations and h = 13).

crystals

MDPI

Editorial

Compounds with Polar Metallic Bonding

Constantin Hoch

Department Chemie, LMU München, Butenandtstraße 5-13(D), D-81377 München, Germany;
constantin.hoch@cup.lmu.de; Tel.: +49-89-2180-77421

Received: 14 May 2019; Accepted: 16 May 2019; Published: 22 May 2019

Recently, I witnessed a discussion amongst solid state chemists whether the term *polar intermetallic bonding* was necessary or dispensable, whether a conceptual discernation of this special class of intermetallic compounds was indicated or spurious. It quickly outcropped that the reason for this discussion is the ambiguity of the term *polar*. Most chemists associate *polarity* immediately with bond polarity in a classical van Arkel-Ketelaar triangle picture [1,2]. And as introduction of ionic polarization into a covalent bond is a very common case also in intermetallic systems, the term *polar intermetallic phases* indeed may seem dispensable. However, the term has existed in the literature for many decades, and there is a good reason for this. Polarity in intermetallic phases causes a number of effects, and the underlying structure-property relationships justify summarizing this class of intermetallic compounds with one common epithet. The conceptual difficulty with it is due to multiple meanings of the term. There are several instances of *polar metal* or *polar metal-metal bonding* in the literature, and as they originate from different scientific backgrounds it is not always clear to the public in which sense *polarity* is being referred to by the author.

Not only Coulombic, but all kinds of dipoles are appropriate to create polarity in an intermetallic phase. The different aspects of macroscopic polarity have one common condition, and it is a crystallographic one. Dipole interaction in a long-range ordering is only observed when inversion symmetry or mirror planes perpendicular to the dipole axis are absent. Therefore the crystallographic meaning of *polar* is the absence of special symmetry operations [3,4]. The perhaps largest number of scientific publications on *polar metals* concentrates on electron conducting materials showing some kind of ordering of electric dipoles in the structures. The coexistence of ordered electric dipoles, as e.g., in ferroelectrics, and metallic behavior comes as a surprise as it would normally be forbidden by Gauss' law: Due to charge screening the effective field within an electron conductor has to be zero, ruling out any kind of cooperative long-range dipole ordering. This rule can be broken in cases of weak electron-phonon coupling, and it is observed in a large and growing number of perovskite-type materials [5–8]. These materials show great potential in future data storage systems with high density and long lifetimes [9,10]. Also the presence of magnetic dipoles and their long-range ordering leads to a form of polarity within an intermetallic phase, and ferromagnetic behavior is a common case. The interface created by contacting a semiconductor with a metal results in a Schottky barrier, and its height depends on electron concentrations, doping and other parameters. The height of the Schottky barrier creates polarity at the metallic interface often referred to in literature as polar bonding [11,12]. And finally, in coordination chemistry, a covalent bonding between the metal centers of a heterodimetallic coordination compound is described as a polar metal-metal bond when the electronegativity differences between the metal atoms is pronounced [13]. This shows how different the meaning of *polar metallic bonding* can be understood, depending on the context.

The Special Issue of *Crystals* entitled *Compounds with Polar Metallic Bonding* presented here is a compilation of eight original articles based on the most recent research projects. It may therefore be seen as a snapshot view on the subject, and it is my great pleasure to see so many different interpretations of

Crystals **2019**, *9*, 267; doi:10.3390/cryst9050267   www.mdpi.com/journal/crystals

the term *polar metallic bonding* assembled here. The broad spectrum of the different meanings of *polarity* in intermetallic compounds is brought forward by a plethora of modern synthetic approaches, structural studies, interpretations of chemical bonding and application-driven materials science. We are extremely happy to have attracted prominent and outstanding members of the intermetallic community to contribute with articles of highest quality to this compilation and we owe them the deepest gratitude:

- Corinna Lorenz, Stefanie Gärtner and Nikolaus Korber report in their article 'Amoniates of Zintl Phases: Similarities and Differences of Binary Phases A_4E_4 and Their Corresponding Solvates' [14] about Zintl chemistry, presenting chemical examples for highest polarity, the complete electron transfer from less noble metal to an electronegative metal. Intermetallic phases of this kind can be dissolved in and recrystallized from polar solvents. Crystalline solvates of Zintl phases may be seen as 'expanded metals' and cross the border from intermetallic phases to coordination compounds in an impressive way.

- Alexander Ovchinnikov, Matej Bobnar, Yurii Prots, Walter Schnelle, Peter Höhn and Yuri Grin present a communication with he title 'Ba$_4$[Mn$_3$N$_6$], a Quasi-One-Dimensional Mixed-Valent Nitridomanganate(II,IV)' [15] and give a beautiful example of both sophisticated modern solid state synthesis and of modern interpretation of the chemical bond in a semiconducting material with long-range ordering of magnetic dipoles. The interplay of magnetic and electronic properties is most interesting in this chain compound.

- Yufei Hu, Kathleen Lee and Susan M. Kauzlarich report on 'Optimization of Ca$_{14}$MgSb$_{11}$ through Chemical Substitutions on Sb Sites: Optimizing Seebeck Coefficient and Resistivity Simultaneously' [16]. Their reseach on thermoelectric materials within the class of Zintl compounds has gained great atention over the years. Getting control over thermal end electric conductivity via structural modification is a highly difficult task, and the article present in this Special Issue gives an excellent example.

- Riccardo Freccero, Pavlo Solokha, Davide Maria Proserpio, Adriana Saccone and Serena De Negri report on 'Lu$_5$Pd$_4$Ge$_8$ and Lu$_3$Pd$_4$Ge$_4$: Two More Germanides among Polar Intermetallics' [17]. Their structural and theoretical study shows the compounds to consist of a network of negatively polarized Ge and Pd atoms whereas Lu acts as a counter-cation, being positively polarized.

- Michael Langenmaier, Michael Jehle and Caroline Röhr present an article entitled 'Mixed Sr and Ba Tri-Stannides/Plumbides A^{II}(Sn$_{1-x}$Pb$_x$)$_3$' [18], dealing with a mixed-crystal series in which the continuous chemical exchange causes the transition from ionic to metallic bonding. This is a most instructive example how chemical bonding can be directly manipulated by chemical means. Modern ways of conceptualizing electron distributions in the sense of counting rules are presented next to high-level DFT calculations of the electronic structures and also geometric analyses.

- Asa Toombs and Gordon J. Miller show a detailed structural study on 'Rhombohedral Distortion of the Cubic MgCu$_2$-Type Structure in Ca$_2$Pt$_3$Ga and Ca$_2$Pd$_3$Ga' [19]. They give an excellent example on how electronic structure and crystallographic distortion mutually interact.

- Fabian Eustermann, Simon Gausebeck, Carsten Dosche, Mareike Haensch, Gunther Wittstock and Oliver Janka present an article entitled 'Crystal Structure, Spectroscopic Investigations, and Physical Properties of the Ternary Intermetallic REPt$_2$Al$_3$ (RE = Y, Dy–Tm) and RE_2Pt$_3$Al$_4$ Representatives (RE = Tm, Lu)' [20]. Here, structural and chemical modifications go hand in hand with symmetry reduction, magnetic interactions and with gradual polarity changes.

- Simon Steinberg and Richard Dronskowski present a review on 'The Crystal Orbital Hamilton Population (COHP) Method as a Tool to Visualize and Analyze Chemical Bonding in Intermetallic Compounds' [21]. This comprehensive study gives a summary and overview on fundamental concepts of recognizing the chemical bonding in intermetallic compounds. They give a coherent introduction into the well-established COHP method, the 25th anniversary of which gave rise for this review. With the examples of cluster-based rare-earth transition metal halides and of gold-containing intermetallic series they illustrate polarity and its expression in terms of bond analyses. The relevance of such considerations on material chemistry is emphasized with respect to phase-change materials and to magnetic materials.

The world of intermetallic compounds with polar metallic bonding is a rapidly growing one. It is a fertile ground on which novel materials emerge, due to the unique ability of polar intermetallics to provide new and unexpected combinations of properties. This Special Issue may be taken as an excellent example on how much further work is needed in order to purposefully direct material research in this field, and, indeed, how valuable basic research on chemical systems and development of concepts for elucidation of electronic bonding situations is with this respect.

References

1. Van Arkel, A.E. *Moleculen en Kristallen*; van Stockum: Den Haag, The Netherlands, 1941.
2. Ketelaar, J.A.A. *De Chemische Binding: Inleiding in de Theoretische Chemie*; Elsevier: New York, NY, USA; Amsterdam, The Netherlands, 1947.
3. Anderson, P.W.; Blount, E.I. Symmetry considerations on martensitic transformations: 'ferroelectric' metals? *Phys. Rev. Lett.* **1965**, *14*, 217. [CrossRef]
4. Lawson, A.C.; Zachariasen, W.H. Low temperature lattice transformation of HfV$_2$. *Phys. Lett.* **1972**, *38*, 1. [CrossRef]
5. Kim, T.H.; Puggioni, D.; Yuan, Y.; Xie, L.; Zhou, H.; Campbell, N.; Ryan, P.J.; Hoi, Y.C.; Kim, J.-W.; Patzner, J.R.; et al. Polar metals by geometric design. *Nature* **2016**, *533*, 68–72. [CrossRef] [PubMed]
6. Puggioni, D.; Rondinelli, J.M. Designing a robustly metallic noncentrosymmetric ruthenate oxide with large thermopower anisotropy. *Nat. Commun.* **2014**, *5*, 3432. [CrossRef] [PubMed]
7. Puggioni, D.; Giovanetti, G.; Capone, M.; Rondinelli, J.M. Design of a Mott multiferroic from a nonmagentic polar metal. *Phys. Rev. Lett.* **2015**, *115*, 087202. [CrossRef] [PubMed]
8. Shi, Y.; Guo, Y.; Wang, X.; Princep, A.J.; Khalyavin, S.; Manuel, P.; Michiue, Y.; Sato, A.; Tsuda, K.; Yu, S.; et al. A ferroelectric-like structural transition in a metal. *Nat. Mater.* **2013**, *12*, 1024–1027. [CrossRef]
9. Scott, J.F. Data storage: Multiferroic memories. *Nat. Mater.* **2007**, *6*, 256–257. [CrossRef]
10. Morin, M.; Canévet, E.; Raynaud, A.; Bartkowiak, M.; Sheptyakov, D.; Ban, V.; Kenzelmann, M.; Pomjakushina, E.; Conder, K.; Medarde, M. Tuning magnetic spirals beyond room temperature with chemical disorder. *Nat. Commun.* **2016**, *7*, 13758. [CrossRef]
11. Mönch, W. (Ed.) *Electronic Structure of Metal-Semiconductor Contacts*; Jaca Book: Milano, Italy, 1990; ISBN 978-94-009-0657-0.
12. Berthold, C.; Binggeli, N.; Baldereschi, A. Schottky barrier heights at polar metal/semiconductor interfaces. *Phys. Rev. B* **2003**, *68*, 085323. [CrossRef]
13. Muetterties, E.L.; Rhodin, T.N.; Band, E.; Brucker, C.F.; Pretzer, W.R. Clusters and Surfaces. *Chem. Rev.* **1979**, *79*, 91–137. [CrossRef]
14. Lorenz, C.; Gärtner, S.; Korber, N. Ammoniates of Zintl phases. similarities and differences of binary phases A$_4$E$_4$ and their corresponding solvates. *Crystals* **2018**, *8*, 276. [CrossRef]
15. Ovchinnikov, A.; Bobnar, M.; Prots, Y.; Schnelle, W.; Höhn, P.; Grin, Y. Ba$_4$[Mn$_3$N$_6$], a quasi-one-dimensional mixed-valent nitridomanganate(II,IV). *Crystals* **2018**, *8*, 235. [CrossRef]
16. Hu, Y.; Lee, K.; Kauzlarich, S.M. Optimization of Ca$_{14}$MgSb$_{11}$ through chemical substitutions on Sb sites: optimizing Seebeck coefficient and resistivity simultaneously. *Crystals* **2018**, *8*, 211. [CrossRef]
17. Freccero, R.; Solokha, P.; Proserpio, D.M.; Saccone, A.; De Negri, S. Lu$_5$Pd$_4$Ge$_8$ and Lu$_3$Pd$_4$Ge$_4$: Two more germanides among polar intermetallics. *Crystals* **2018**, *8*, 205. [CrossRef]

18. Langenmaier, M.; Jehle, M.; Röhr, C. Mixed Sr and Ba tri-stannides/plumbides $A^{II}(Sn_{1-x}Pb_x)_3$. *Crystals* **2018**, *8*, 204. [CrossRef]
19. Toombs, A.; Miller, G.J. Rhombohedral distortion of the cubic $MgCu_2$-type structure in Ca_2Pt_3Ga and Ca_2Pd_3Ga. *Crystals* **2018**, *8*, 186. [CrossRef]
20. Eustermann, F.; Gausebeck, S.; Dosche, C.; Haensch, M.; Wittstock, G.; Janka, O. Crystal structure, spectroscopic investigations, and physical properties of the ternary intermetallic $REPt_2Al_3$ (RE = Y, Dy-Tm) and $RE_2Pt_3Al_4$ representatives (RE = Tm, Lu). *Crystals* **2018**, *8*, 169. [CrossRef]
21. Steinberg, S.; Dronskowski, R. The crystal orbital Hamilton population (COHP) method as a tool to visualize and analyze chemical bonding in intermetallic compounds. *Crystals* **2018**, *8*, 225. [CrossRef]

crystals

MDPI

Article

Ammoniates of Zintl Phases: Similarities and Differences of Binary Phases A$_4$E$_4$ and Their Corresponding Solvates

Corinna Lorenz, Stefanie Gärtner and Nikolaus Korber *

Institute of Inorganic Chemistry, University of Regensburg, 93055 Regensburg, Germany;
Corinna.Lorenz@ur.de (C.L.); Stefanie.Gaertner@ur.de (S.G.)
* Correspondence: Nikolaus.Korber@ur.de; Tel: +49-941-943-4448; Fax: +49-941-943-1812

Received: 20 April 2018; Accepted: 23 June 2018; Published: 29 June 2018

Abstract: The combination of electropositive alkali metals A (A = Na-Cs) and group 14 elements E (E = Si-Pb) in a stoichiometric ratio of 1:1 in solid state reactions results in the formation of polyanionic salts, which belong to a class of intermetallics for which the term Zintl compounds is used. Crystal structure analysis of these intermetallic phases proved the presence of tetrahedral tetrelide tetraanions [E$_4$]$^{4-}$ precast in solid state, and coulombic interactions account for the formation of a dense, three-dimensional cation-anion network. In addition, it has been shown that [E$_4$]$^{4-}$ polyanions are also present in solutions of liquid ammonia prepared via different synthetic routes. From these solutions crystallize ammoniates of the alkali metal tetrahedranides, which contain ammonia molecules of crystallization, and which can be characterized by X-ray crystallography despite their low thermal stability. The question to be answered is about the structural relations between the analogous compounds in solid state vs. solvate structures, which all include the tetrahedral [E$_4$]$^{4-}$ anions. We here investigate the similarities and differences regarding the coordination spheres of these anions and the resulting cation-anion network. The reported solvates Na$_4$Sn$_4$·13NH$_3$, Rb$_4$Sn$_4$·2NH$_3$, Cs$_4$Sn$_4$·2NH$_3$, Rb$_4$Pb$_4$·2NH$_3$ as well as the up to now unpublished crystal structures of the new compounds Cs$_4$Si$_4$·7NH$_3$, Cs$_4$Ge$_4$·9NH$_3$, [Li(NH$_3$)$_4$]$_4$Sn$_4$·4NH$_3$, Na$_4$Sn$_4$·11.5NH$_3$ and Cs$_4$Pb$_4$·5NH$_3$ are considered for comparisons. Additionally, the influence of the presence of another anion on the overall crystal structure is discussed by using the example of a hydroxide co-crystal which was observed in the new compound K$_{4.5}$Sn$_4$(OH)$_{0.5}$·1.75 NH$_3$.

Keywords: Zintl compounds; liquid ammonia; crystal structure

1. Introduction

The term "polar intermetallics" applies to a large field of intermetallic compounds, the properties of which range from metallic and superconducting to semiconducting with a real band gap [1–5]. For the compounds showing a real band gap, the Zintl–Klemm concept is applicable by formally transferring the valence electrons of the electropositive element to the electronegative partner, and the resulting salt-like structure allows for the discussion of anionic substructures [1–9]. The combination of electropositive alkali metals A (A = Na-Cs) and group 14 elements E (E = Si-Pb) in a stoichiometric ratio of 1:1 in solid state reactions results in the formation of salt-like, semiconducting intermetallic compounds which show the presence of the tetrahedral [E$_4$]$^{4-}$ anions precast in solid state. These anions are valence isoelectronic to white phosphorus and can be seen as molecular units. They have been known since the work of Marsh and Shoemaker in 1953 who first reported on the crystal structure of NaPb [10]. Subsequently, the list of the related binary phases of alkali metal and group 14 elements was completed (Table 1, Figure 1). Due to coulombic interactions a dense, three-dimensional cation-anion network in either the KGe structure type (A = K-Cs; E = Si, Ge) [11–17] or NaPb structure type (A = Na-Cs; E = Sn, Pb) (Figure 1e,f) [18–21] is observed. For sodium and the lighter group 14 elements silicon and germanium, binary compounds lower in symmetry (NaSi: *C*2/*c* [14,16,17,22], NaGe: *P*2₁/*c* [14,16]) are formed, which also contain the tetrahedral shaped [E₄]⁴⁻ anions (Figure 1c,d). In the case of lithium, no binary compound with isolated [E₄]⁴⁻ polyanions is reported at ambient conditions: In LiSi [23] and LiGe [24] (LiSi structure type, Figure 1a), threefold bound silicon atoms are observed in a three-dimensionally extended network, which for tetrel atoms with a charge of −1 is an expected topological alternative to tetrahedral molecular units, and which conforms to the Zintl–Klemm concept. If the [E₄]⁴⁻ cages are viewed as approximately spherical, the calculated radius *r* would be 3.58 Å for silicide, 3.67 Å for germanide, 3.96 Å for stannide and 3.90 Å for plumbide clusters (*r* = averaged distances of the center of the cages to the vertex atoms + van der Waals radii of the elements, each [25]). The dimensions for silicon and germanium are very similar, as are those for tin and lead. It is worth noticing that there is a significant increase in the size of the tetrahedra, which are considered as spherical, for the transition from germanium to tin, which could explain the change of the structure type KGe to NaPb.

For binary compounds of lithium and tin or lead, the case is different. The Zintl rule is not applicable as LiSn (Figure 1b) [26] includes one-dimensional chains of tin atoms, whereas NaSn [19,27] forms two-dimensional layers as the tin substructure. For LiPb [28] the CsCl structure has been reported, which is up to now unreproduced.

Table 1. Binary phases of alkali metal (Li-Cs) and group 14 element with 1:1 stoichiometric ratio (ambient conditions).

	Si	Ge	Sn	Pb
Li	*I*4₁/*a* LiSi [23]	*I*4₁/*a* LiSi [24]	*I*4₁/*amd* [26]	CsCl (?) [28]
Na	*C*2/*c* [14,16,17,22]	*P*2₁/*c* [14,16]	*I*4₁/*acd* NaPb [19,27]	*I*4₁/*acd* NaPb [10]
K	*P*-43*n* KGe [11,12,14]	*P*-43*n* KGe [11,13]	*I*4₁/*acd* NaPb [19,21]	*I*4₁/*acd* NaPb [21]
Rb	*P*-43*n* KGe [11,12,14]	*P*-43*n* KGe [11,13,14]	*I*4₁/*acd* NaPb [18,21]	*I*4₁/*acd* NaPb [20,21]
Cs	*P*-43*n* KGe [11,12,14]	*P*-43*n* KGe [11,13,14]	*I*4₁/*acd* NaPb [18,21]	*I*4₁/*acd* NaPb [24]

Additionally, it has been shown that the tetrelide tetraanions are also present in solutions of liquid ammonia [29], and from these solutions alkali metal cation-[E₄]⁴⁻ compounds that additionally contain ammonia molecules of crystallization can be precipitated. We earlier reported on the crystal structures of Rb₄Sn₄·2NH₃, Cs₄Sn₄·2NH₃ and Rb₄Pb₄·2NH₃, which showed strong relations to the corresponding binaries [30]. In Na₄Sn₄·13 NH₃ [31,32] no such relation is observed. In general, ammonia in solid ammoniates is not only an innocent and largely unconnected solvent molecule but may also act as a ligand towards the alkali metal cations. This leads to a variety of crystal structures, which allows for the investigation of the competing effects of cation-anion-interaction vs. alkali-metal-ammine complex formation in the solid state. We here report on the single crystal X-ray investigations of the new compounds Cs₄Si₄·7NH₃, Cs₄Ge₄·9NH₃, [Li(NH₃)₄]₄Sn₄·4NH₃, Na₄Sn₄·11.5NH₃ and Cs₄Pb₄·5NH₃ and compare the previously reported solvates as well as the new ammoniate compounds of tetratetrelide tetranions to the known binary compounds. It has to be noted that the number of ammoniate structures of tetrelide tetraanions is very limited [30–34] as they are easily oxidized in solution by forming less reduced species like [E₉]⁴⁻ [35–40] and [E₅]²⁻ [36,41–43]. In Table 2, all hitherto known ammoniates which contain the highly charged [E₄]⁴⁻ (E = Si-Pb) cluster are listed. For [Sn₉]⁴⁻ we could recently show that co-crystallization of hydroxide anions is possible in the compound Cs₅Sn₉(OH)·4NH₃ [44]. We here present the first crystal structure of the co-crystal of [Sn₄]⁴⁻ and the hydroxide anion in the compound K₄.₅Sn₄(OH)₀.₅·1.75 NH₃ which allows for the discussion of the influence of another anion on the overall crystal structure.

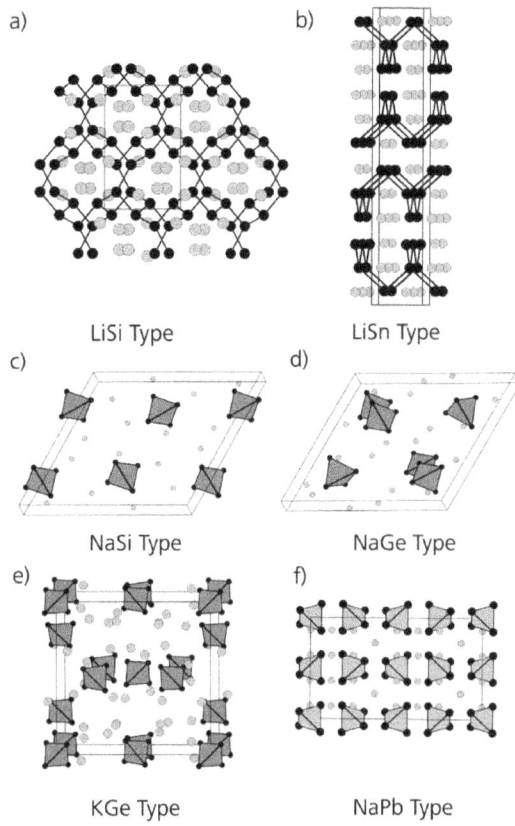

Figure 1. Different structure types for the binary phases of alkali metal (Li-Cs) and group 14 element with 1:1 stoichiometric ratio (ambient conditions) (**a**–**f**).

Table 2. Hitherto known $A_4E_4 \cdot x NH_3$ (A = Li-Cs; E = Si-Pb) solvate structures and selected crystal structure details. The bold marked new compounds are discussed in this article.

	Compound	Crystal System	Space Group	Unit Cell Dimensions
Si	**Cs$_4$Si$_4$·7NH$_3$**	triclinic	P-1	a = 12.3117(6) Å; b = 13.0731(7) Å; c = 13.5149(7) Å; V = 2035.88(19) Å^3
Ge	**Cs$_4$Ge$_4$·9NH$_3$**	orthorhombic	*Ibam*	a = 11.295(2) Å; b = 11.6429(15) Å; c = 17.237(2) Å; V = 2266.9(6) Å^3
Sn	**[Li(NH$_3$)$_4$]$_4$Sn$_4$·4NH$_3$**	monoclinic	$I2/a$	a = 16.272(3) Å; b = 10.590(2) Å; c = 20.699(4) Å; V = 3446.9(13) Å^3
	[Li(NH$_3$)$_4$]$_9$Li$_3$(Sn$_4$)$_3$·11NH$_3$ [31]	monoclinic	$P2/n$	a = 12.4308(7) Å; b = 9.3539(4) Å; c = 37.502(2) Å; V = 4360.4(4) Å^3
	Na$_4$Sn$_4$·11.5NH$_3$	monoclinic	$P2_1/c$	a = 13.100(3) Å; b = 31.393(6) Å; c = 12.367(3) Å; V = 5085.8(18) Å^3
	Na$_4$Sn$_4$·13NH$_3$ [31,32]	hexagonal	$P6_3/m$	a = b = 10.5623(4) Å; c = 29.6365(16) Å; V = 2863.35 Å^3
	K$_4$Sn$_4$·8NH$_3$ [31]	hexagonal	$P6_3$	a = b = 13.1209(4) Å; c = 39.285(2) Å; V = 5857.1(4) Å^3
	Rb$_4$Sn$_4$·2NH$_3$ [30]	monoclinic	$P2_1/a$	a = 13.097(4) Å; b = 9.335(2) Å; c = 13.237(4) Å; V = 1542.3(7) Å^3
	Cs$_4$Sn$_4$·2NH$_3$ [30]	monoclinic	$P2_1/a$	a = 13.669(2) Å; b = 9.627(1) Å; c = 13.852(2) Å; V = 1737.6(4) Å^3
Pb	Rb$_4$Pb$_4$·2NH$_3$ [30]	monoclinic	$P2_1/a$	a = 13.170(3) Å; b = 9.490(2) Å; c = 13.410(3) Å; V = 1595.2(6) Å^3
	Cs$_4$Pb$_4$·5NH$_3$	orthorhombic	*Pbcm*	a = 9.4149(3) Å; b = 27.1896(7) Å; c = 8.1435(2) Å; V = 2084.63(10) Å^3

2. Materials and Methods

For the preparation of $[E_4]^{4-}$-containing solutions different preparative routes are possible which are described elsewhere [8]. In general, liquid ammonia was stored over sodium metal and was directly condensed on the reaction mixture under inert conditions (see Appendix A). The reaction vessels were stored for at least three months at 235 K or 197 K. For the handling of the very temperature and moisture labile crystals, a technique developed by Kottke and Stalke was used [45,46]. Crystals were isolated directly with a micro spatula from the reaction solutions in a recess of a glass slide containing perfluoroether oil, which was cooled by a steam of liquid nitrogen. By means of a stereo microscope, an appropriate crystal was selected and subsequently attached on a MicroLoop™ and placed on a goniometer head on the diffractometer. For details on the single crystal X-Ray structure analysis, please see Table 3.

Table 3. Crystal structure and structure refinement details for the compounds described above.

Chemical Formula	Cs4Pb4·5NH3	Cs4Ge4·9NH3	Cs4Si4·7NH3	Na4Sn4·11.5NH3	[Li(NH3)4]4Sn4·4NH3	K4.5Sn4(OH)0.5·1.75NH3
CSD No. *	434173	434172	434176	421860	421857	427472
M_r [g·mol^{-1}]	1445.57	948.09	763.24	1525.25	843.20	689.03
Crystal system	orthorhombic	orthorhombic	triclinic	monoclinic	monoclinic	monoclinic
Space group	$Pbcm$	$Ibam$	$P\text{-}1$	$P2_1/c$	$I2/a$	$P2_1/c$
a [Å]	9.4149(3)	11.295(2)	12.3117(6)	13.100(3)	16.272(3)	16.775(3)
b [Å]	27.1896(7)	11.6429(15)	13.0731(7)	31.393(6)	10.590(2)	13.712(3)
c [Å]	8.1435(2)	17.237(2)	13.5149(7)	12.367(3)	20.699(4)	26.038(5)
α [°]	90	90	85.067(4)	90	90	90
β [°]	90	90	73.052(4)	90.32(3)	104.90(3)	90.92(3)
γ [°]	90	90	78.183(4)	90	90	90
V [Å^3]	2084.63(10)	2266.9(6)	2035.88(19)	5085.8(18)	3446.9(13)	5988(2)
Z	4	4	4	4	4	16
$F(000)$ (e)	2392.0	1644.0	1384.0	2800.0	1648.0	4920.0
ρ_{calc} [g·cm^{-3}]	4.606	2.778	2.490	1.968	1.625	3.057
μ [mm^{-1}]	39.072	11.578	7.331	3.955	2.887	7.807
Absorption correction	numerical [47]	/	numerical [47]	numerical [48]	numerical [48]	numerical [48]
Diffractometer (radiation source)	Super Nova (Mo)	Super Nova (Mo)	Super Nova (Mo)	Stoe IPDS II (Mo)	Stoe IPDS II (Mo)	Stoe IPDS II (Mo)
2θ- range for data collection [°]	6.24–52.74	6.9–48.626	6.3–50.146	3.892–51.078	4.072–50.91	3.836–50.966
Reflections collected/independent	18834/2274	2294/748	26514/7197	9587/9390	22976/3118	27272/10460
Data/restraints/parameters	2274/0/72	748/0/44	7197/30/377	9390/0/370	3118/9/163	10460/0/389
Goodness-of-fit on F^2	1.086	1.043	1.038	0.802	0.886	0.844
Final R indices [$I > 2\sigma(I)$]	R1 = 0.0388, wR2 = 0.0900	R1 = 0.0711, wR2 = 0.1251	R1 = 0.0304, wR2 = 0.0747	R1 = 0.0401, wR2 = 0.1007	R1 = 0.0400, wR2 = 0.0798	R1 = 0.0592, wR2 = 0.1397
R indices (all data)	R1 = 0.0425, wR2 = 0.0926	R1 = 0.1323, wR2 = 0.1525	R1 = 0.0365, wR2 = 0.0780	R1 = 0.0625, wR2 = 0.1101	R1 = 0.0748, wR2 = 0.0861	R1 = 0.1037, wR2 = 0.1538
R_{int}	0.0884	0.1162	0.0343	0.1011	0.0965	0.0704
$\Delta\rho_{max}$, $\Delta\rho_{min}$ [e·Å^{-3}]	2.48/−2.32	1.70/−1.24	2.00/−2.22	1.90/−1.17	1.61/−0.62	3.86/−1.24

* Further details of the crystal structure investigations may be obtained from FIZ Karlsruhe, 76344 Eggenstein-Leopoldshafen, Germany (Fax: (+49)7247-808-666; e-mail: crysdata(at)fiz-karlsruhe(dot)de, on quoting the deposition numbers.

3. Results

In the following, the crystal structures of the new compounds $Cs_4Pb_4 \cdot 5NH_3$, $Cs_4Ge_4 \cdot 9NH_3$, $Cs_4Si_4 \cdot 7NH_3$, $Na_4Sn_4 \cdot 11.5NH_3$, $[Li(NH_3)_4]_4Sn_4 \cdot 4NH_3$ and $K_{4.5}Sn_4(OH)_{0.5} \cdot 1.75NH_3$ are described independently, their similarities and differences towards the binary materials are discussed subsequently in Section 4 (Discussions).

3.1. $Cs_4Pb_4 \cdot 5NH_3$

The reaction of elemental lead with stoichiometric amounts of cesium in liquid ammonia yields shiny metallic, reddish needles of $Cs_4Pb_4 \cdot 5NH_3$. The asymmetric unit of the crystal structure of $Cs_4Pb_4 \cdot 5NH_3$ consists of three crystallographically independent lead atoms, four cesium cations and four ammonia molecules of crystallization. One of the lead atoms and one of the nitrogen atoms are located on the general *Wyckoff* position *8e* of the orthorhombic space group *Pbcm* (No. 57). The other two lead atoms, four Cs$^+$ cations and three nitrogen atoms occupy the special *Wyckoff* positions *4d* (mirror plane) and *4c* (twofold screw axis) with a site occupancy factor of 0.5 each. The Pb$_4$ cage is generated from the three lead atoms through symmetry operations. As there is no structural indication for the ammonia molecules to be deprotonated, the $[Pb_4]^{4-}$ cage is assigned a fourfold negative charge, which is compensated by the four cesium cations. The Pb-Pb distances within the cage range between 3.0523(7) Å and 3.0945(5) Å. They are very similar to those that have been found in the solventless binary structures (3.090(2) Å) [21]. The cluster has a nearly perfect tetrahedral shape with angles close to 60°. The tetraplumbide tetraanion is coordinated by twelve Cs$^+$ cations at distances between 3.9415(1)–5.4997(8) Å. They coordinate edges, faces and vertices of the cage (Figure 2e). The coordination sphere of Cs1 is built up by four $[Pb_4]^{4-}$ cages (3 × η^1, 1 × η^2) and five ammonia molecules of crystallization. Here, the cesium cation is surrounded by four lead clusters tetrahedrally and thus forms a supertetrahedron (Figure 3a).

Cs2 and Cs4 are trigonally surrounded by three Pb$_4$ cages (1 × η^1, 2 × η^2 and 2 × η^1, 1 × η^3) each. Their coordination spheres are completed by five and four ammonia molecules of crystallization, respectively, as shown for Cs2 in Figure 3b. Cs3 only shows contacts to two Pb$_4$ cages (2 × η^2) and six ammonia molecules of crystallization (Figure 3c). Altogether, a two-dimensional network is formed. Along the crystallographic b-axis, corrugated Cs$^+$-NH$_3$ strands are built. The $[Pb_4]^{4-}$ cages are situated along the strands and are stacked along the c-axis (Figure 4).

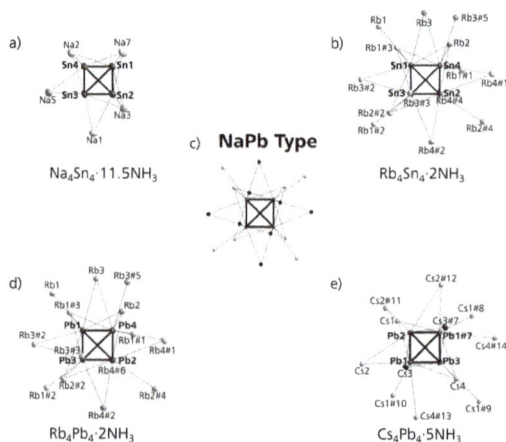

Figure 2. Comparison of the cationic coordination spheres of $[E_4]^{4-}$ (E = Sn, Pb) clusters in $Na_4Sn_4 \cdot 11.5NH_3$ (**a**); $Rb_4Sn_4 \cdot 2NH_3$ (**b**); NaPb (**c**); $Rb_4Pb_4 \cdot 2NH_3$ (**d**) and $Cs_4Pb_4 \cdot 5NH_3$ (**e**); probability factor: 50%; dark grey marked cations occupy special *Wyckoff* positions.

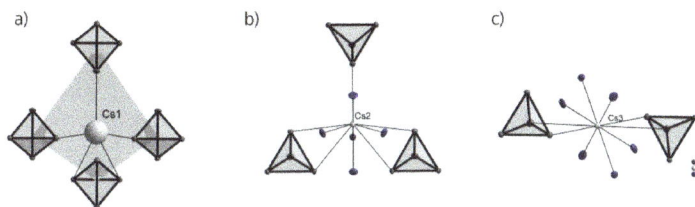

Figure 3. Coordination spheres of the cesium cations in $Cs_4Pb_4 \cdot 5NH_3$; (**a**) tetrahedral environment of Cs1 by $[Pb_4]^{4-}$, for reasons of clarity, ammonia molecules are omitted; (**b,c**) coordination spheres of Cs2 (representative for Cs4) and Cs3; for reasons of clarity, hydrogen atoms are omitted; probability factor: 50%.

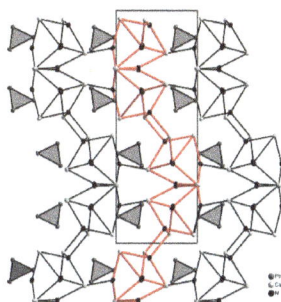

Figure 4. Section of the structure of $Cs_4Pb_4 \cdot 5NH_3$; corrugated Cs^+-NH_3 strands along the crystallographic b-axis; the chains are emphasized by bold lines; $[Pb_4]^{4-}$ cages are located along the strands; hydrogen atoms are omitted for clarity; probability factor: 79%.

3.2. $Cs_4Ge_4 \cdot 9NH_3$

Deep red needles of $Cs_4Ge_4 \cdot 9NH_3$ could be obtained by the dissolution of $Cs_{12}Ge_{17}$ together with two chelating agents, [18]crown-6 and [2.2.2]cryptand in liquid ammonia. Indexing of the collected reflections leads to the orthorhombic space group *Ibam* (No. 72). The asymmetric unit of this compound consists of one germanium atom, one cesium cation and four nitrogen atoms. The anionic part of the compound is represented by a $[Ge_4]^{4-}$ tetrahedron, which is generated by the germanium position through symmetry operations resulting in the point group D_2 for the molecular unit. The definite number of ammonia molecules of crystallization cannot be determined due to the incomplete data set (78%), but very likely sums up to four in the asymmetric unit. $Cs_4Ge_4 \cdot 9NH_3$ is the first ammoniate with a ligand-free tetragermanide tetraanion reported to date. In spite of the incomplete data set, the heavy atoms Cs and Ge could be unambiguously assigned as maxima in the Fourier difference map. The dimensions of the germanium cage (2.525(3)–2.592(3) Å) comply with the expected values found in literature (2.59 Å [11]). The $[Ge_4]^{4-}$ anion shows almost perfect tetrahedral symmetry with Ge-Ge-Ge angles between 58.63(10)° and 61.21(11)°. It is surrounded by eight cesium cations. They coordinate η^1-like to edges and η^3-like to triangular faces of the cage (Figure 5e). The coordination sphere of the cesium atom itself is built by two $[Ge_4]^{4-}$ cages and is completed by eight ammonia molecules of crystallization. Considering the Cs^+-$[Ge_4]^{4-}$ contacts, layers parallel to the crystallographic a- and b-axis are formed, which are separated by ammonia molecules of crystallization.

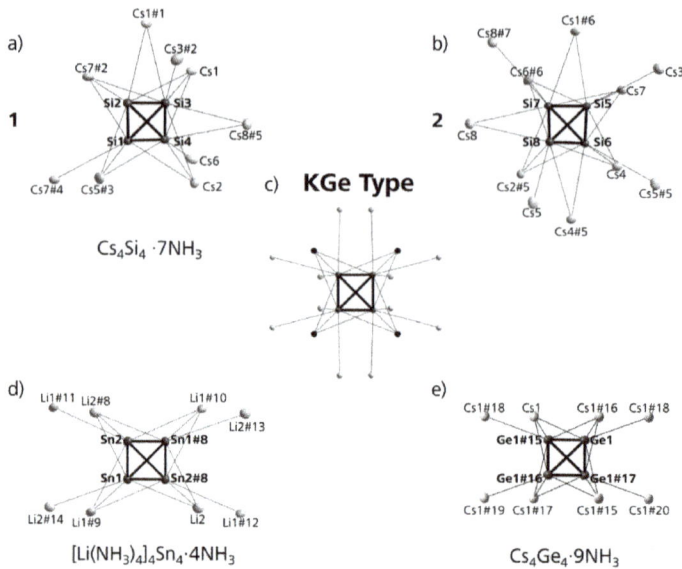

Figure 5. Comparison of the cationic coordination spheres of $[E_4]^{4-}$ (E = Si-Sn) clusters in $Cs_4Si_4 \cdot 7NH_3$ (**a,b** (two crystallographically independent $[Si_4]^{4-}$ cages)), KGe (**c**); $[Li(NH_3)_4]_4Sn_4 \cdot 4NH_3$ (**d**) and $Cs_4Ge_4 \cdot 9NH_3$ (**e**); probability factor: 50%; dark grey marked cations occupy special *Wyckoff* positions.

3.3. $Cs_4Si_4 \cdot 7NH_3$

Dissolving $Cs_{12}Si_{17}$ together with dicyclohexano[18]crown-6 and [2.2.2]cryptand in liquid ammonia resulted in deep red prismatic crystals of $Cs_4Si_4 \cdot 7NH_3$. The asymmetric unit consists of two crystallographically independent $[Si_4]^{4-}$ clusters, eight cesium cations and 14 ammonia molecules of crystallization. All atoms are located on the general *Wyckoff* position *2i* of the triclinic space group *P*-1 (No. 2). $[Si_4]^{4-}$ (**1**) is surrounded by nine, $[Si_4]^{4-}$ (**2**) by eleven Cs^+ cations (Figure 5a,b). Here the cations span edges, faces and vertices of the clusters in a distance range of 3.559(2)–4.651(3) Å. Cs1, Cs5, Cs7 and Cs8 are η^1-, η^2- and η^3-like surrounded by three Si_4 cages each, which are arranged in a triangular shape (comparable to the coordination sphere shown in Figure 3b). The remaining four cesium cations per asymmetric unit also show ionic contacts to two silicon clusters each by spanning edges, faces and vertices of the latter. The coordination spheres of all alkali metal cations are completed by four to nine ammonia molecules of crystallization (Figure 6b). Altogether, a two-dimensional $[Si_4]^{4-}$-Cs^+-network is formed. The anionic cluster and the cations built corrugated waves, comparable to $Cs_4Pb_4 \cdot 5NH_3$ (Figure 4). The ammonia molecules of crystallization fill the space between the strands.

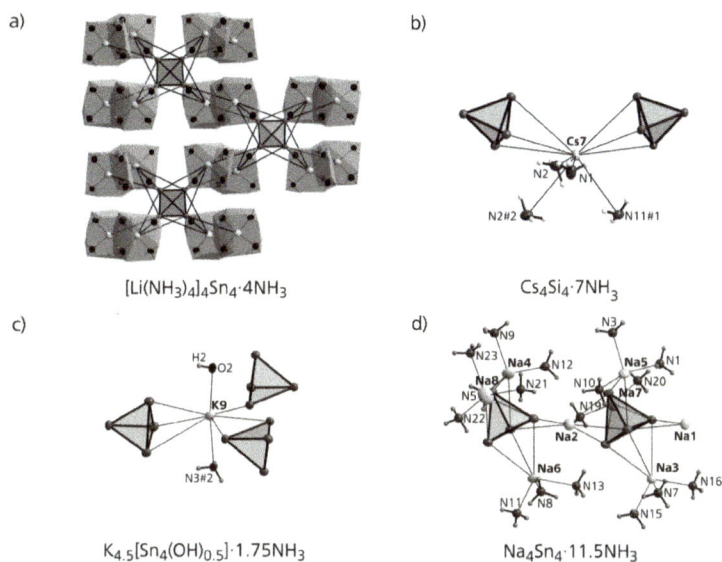

Figure 6. Coordination spheres of the cations; (**a**) $[Li(NH_3)_4]^+_4[Sn_4]^{4-}$ strands; for reasons of clarity, the $[Li(NH_3)_4]^+$ complexes are shown as spherical polyhedra; (**b**) of Cs7 in $Cs_4Si_4 \cdot 7NH_3$, representative for the coordination spheres of the heavier alkali metals; (**c**) of K9 in $K_{4.5}Sn_4(OH)_{0.5} \cdot 1.75NH_3$ as a representative of the other cations in the structure; (**d**) complete coordination sphere of $[Sn_4]^{4-}$ anions and sodium cations; probability factor: 50%.

3.4. $Na_4Sn_4 \cdot 11.5NH_3$

Red, prism-shaped crystals of $Na_4Sn_4 \cdot 11.5NH_3$ could be synthesized by reacting elemental tin with stoichiometric amounts of sodium and ^tBuOH in liquid ammonia. Two crystallographically independent $[Sn_4]^{4-}$ tetrahedra represent the anionic part of the asymmetric unit. The charge is compensated by eight sodium cations. Additionally, 23 ammonia molecules of crystallization can be found. All atoms occupy the general *Wyckoff* position *4e* of the monoclinic space group $P2_1/c$ (No 14). Although the two $[Sn_4]^{4-}$ cages are crystallographically independent, the chemical environment is very similar (Figure 2a). Five sodium cations reside on edges and triangular faces of each cluster. Considering the anion-cation contacts, one-dimensional strands along the a-axis are formed. The tetrastannide clusters are bridged by two crystallographically independent sodium cations Na1 and Na2, which alternatingly coordinate faces and edges of the cages (Figure 6d). Thus the anionic part of the structure can be assigned the formula $_1^\infty[Na(Sn_4)]^{3-}$. A similar coordination of the bridging atom was recently found in the ammoniate $Rb_6[(\eta^2-Sn_4)Zn(\eta^3-Sn_4)] \cdot 5NH_3$, where two $[Sn_4]^{4-}$ anions are bridged by a Zn^{2+} cation forming isolated dimeric units [49]. As already mentioned, Na1 and Na2 only show contacts to $[Sn_4]^{4-}$, the remaining six sodium cations additionally coordinate to ammonia molecules of crystallization. Altogether, a molecular formula of $[(Sn_4)Na]_2[(Na(NH_3)_3)_5(Na(NH_3)_2)] \cdot 6NH_3$ represents the whole crystal structure, where sodium-$[Sn_4]^{4-}$ strands are separated by both coordinating ammonia and unattached ammonia molecules of crystallization.

3.5. $[Li(NH_3)_4]_4Sn_4 \cdot 4NH_3$

The reaction of elemental tin with stoichiometric amounts of lithium and ^tBuOH in liquid ammonia yields in black shaped crystals of $[Li(NH_3)_4]_4Sn_4 \cdot 4NH_3$. The asymmetric unit of the new compound consists of two tin atoms, two lithium atoms and ten ammonia molecules of crystallization.

All atoms are located on the general *Wyckoff* position *8f* of the monoclinic space group *I2/a* (No. 15). As there is no indication for the presence of deprotonated ammonia molecules, the charge of the Sn_4 cluster sums up to -4. The tin cluster does not show direct contacts to lithium cations as all of these are coordinated by four ammonia molecules which results in tetrahedrally shaped $[Li(NH_3)_4]^+$ complexes that can be considered as large and approximately spherical cationic units (Figure 6). For details, see Section 4.2. The ammonia lithium distances of 2.064(1)–2.116(1) Å in the tetrahedral cationic complex $[Li(NH_3)_4]^+$ are in good agreement with literature-known lithiumtetraammine complexes [38]. The $[Sn_4]^{4-}$ cage is coordinated by eight $[Li(NH_3)_4]^+$ complexes, which span vertices and faces of the cluster (Figure 6a). The distances within the tetrahedron vary between 2.9277(8)–2.9417(8) Å and lie within the expected values for $[Sn_4]^{4-}$ anions in ammoniate crystal structures. The Sn-Sn-Sn angles range between 59.882(20)° and 60.276(20)°.

3.6. $K_{4.5}Sn_4(OH)_{0.5} \cdot 1.75NH_3$

Red, prismatic crystals of the composition $K_{4.5}Sn_4(OH)_{0.5} \cdot 1.75NH_3$ could be synthesized by dissolving elemental tin with stoichiometric amounts of potassium in the presence of ^tBuOH in liquid ammonia. The hydroxide in the solvate structure is probably formed due to impurities on the potassium. The asymmetric unit of the solvate structure consists of four tetrastannide tetraanions, two hydroxide ions and seven ammonia molecules of crystallization. They all occupy general *Wyckoff* positions of the monoclinic space group *P2₁/c* (No. 14). The bond lengths of the tetrastannides of 2.884(2)–2.963(2) Å are within the expected values for Sn-Sn distances in tin tetrahedranides [30]. Three of the four crystallographically independent tin anions are coordinated by 14 potassium cations, the fourth anion is coordinated by 16 cations at distances between 3.438(5) Å and 3.145(5) Å (Figure 7a,b). The cations coordinate vertex tin atoms or span edges and faces of the tetrahedra.

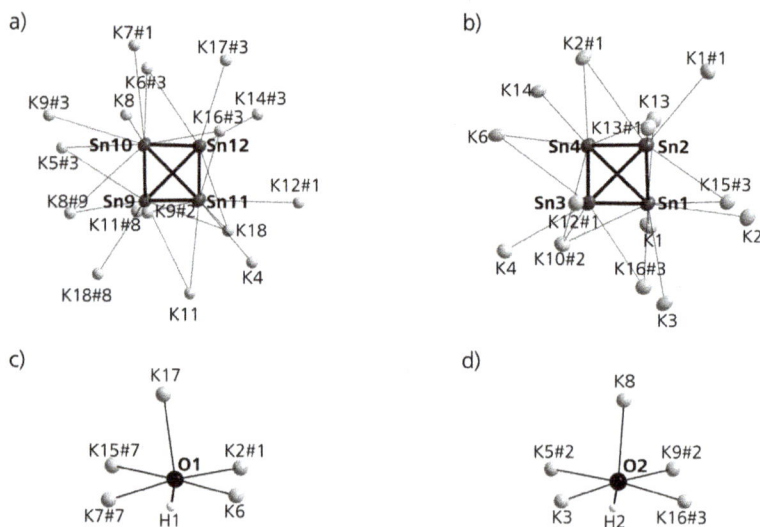

Figure 7. Cationic coordination spheres of the two anionic components in $K_{4.5}Sn_4(OH)_{0.5} \cdot 1.75NH_3$; (**a**) $[Sn_4]^{4-}$ sourrounded by 16 cations; (**b**) $[Sn_4]^{4-}$ coordinated by 14 cations, as a represanttive for the other two crystallographically independent $[Sn_4]^{4-}$ cages in the asymmetric unit; (**c,d**) cationic environment of the two hydoxide anions; probability factor: 50%.

Figure 7 additionally shows the coordination sphere of the second anionic component of the solvate structure, the hydroxide anions. They are characteristically surrounded by five potassium cations in a distorted square pyramidal manner. The coordination sphere of the cations is completed by tin clusters, hydroxide ions or/and ammonia molecules of crystallization (Figure 7c,d). Altogether, the structure of $K_{4.5}Sn_4(OH)_{0.5} \cdot 1.75NH_3$ consists of strands of ammonia molecules, hydroxide anions and potassium cations, which are connected via tetrastannide anions.

4. Discussion

In this section we discuss similarities and differences of the binary compounds towards the solvate structures with respect to the coordination spheres of the cations and the cluster anions.

4.1. NaPb Type Analogies

As already mentioned in the introduction, all alkali metal stannides and plumbides with the nominal composition AE, except the compounds containing lithium, crystallize in the tetragonal space group $I4_1/acd$ (No. 142) and belong to the NaPb structure type [10,18–21]. Considering the direct cationic environment of the tetrelide cluster in the binary phase (Figure 2c), the coordination number (CN) sums up to 16. With increasing content of ammonia molecules of crystallization, the coordination number of the cages decrease (Table 4). Figure 2 shows which cation-anion contacts are broken within the solvate structures. Generally, there are three different modes of the coordination of the cation towards the anion (Figure 8).

In the binary phases and $Na_4Sn_4 \cdot 13NH_3$ [31,32] all triangular faces of the anions are capped η^3-like by cations. In contrast, in $A_4E_4 \cdot 2NH_3$ (A = K, Rb; E = Sn, Pb) [30] three faces and in $Cs_4Pb_4 \cdot 5NH_3$ only one face of the $[E_4]^{4-}$ anions are coordinated η^3-like by the cations. In addition to the coordination of the faces, the edges of the $[E_4]^{4-}$ tetrahedra are coordinated η^2-like. For NaPb, $Rb_4Sn_4 \cdot 2NH_3$, $Cs_4Sn_4 \cdot 2NH_3$ and $Rb_4Pb_4 \cdot 2NH_3$, four η^2-like coordinated cations are present. In $Cs_4Pb_4 \cdot 5NH_3$ five cations coordinate to the cage in a η^2-like fashion, in $Na_4Sn_4 \cdot 11.5NH_3$ only two. Finally, the cationic environment of the $[E_4]^{4-}$ anions in the binary phase is completed by a total of eight cations which are bonded η^1-like to each vertex. In $Rb_4Sn_4 \cdot 2NH_3$, $Cs_4Sn_4 \cdot 2NH_3$, $Rb_4Pb_4 \cdot 2NH_3$ and $Cs_4Pb_4 \cdot 5NH_3$ three and two vertices are coordinated by two cations, respectively. The other vertices each only show one tetrelide-alkali metal contact. Table 4 summarizes the anion coordinations and it becomes evident that the solvate structures with a small content of ammonia molecules of crystallization are more similar to the solid state structure, thus the three-dimensional cation-anion interactions are considerably less disturbed. Additionally, more anion-cation contacts appear in the solvate structures with the heavier alkali metals. Rubidium and cesium, as well as tin and lead are considered as soft acids and bases according to the HSAB theory [50]. The solvate structures containing sodium show much less anion-cation contacts due to the favored interaction of the hard base ammonia to the hard acid sodium cation (Table 5). Table 5 additionally shows the total coordination numbers of the cations, which is classified into cation-anion (A^+-E^-) and cation-nitrogen (A^+-NH_3) contacts. In $Na_4Sn_4 \cdot 13NH_3$, $Na_4Sn_4 \cdot 11.5NH_3$ and $Cs_4Pb_4 \cdot 5NH_3$ the numbers of anion-cation contacts and the cation-nitrogen contacts are similar. In contrast, $Rb_4Sn_4 \cdot 2NH_3$, $Cs_4Sn_4 \cdot 2NH_3$ and $Rb_4Pb_4 \cdot 2NH_3$ show more A^+-E^- contacts than ion-dipole interactions between the cation and the ammonia molecules of crystallization.

Table 4. Coordination number of the $[E_4]^{4-}$ cages in NaPb and related compounds.

Compound	Coordination Number (CN) E^--A^+	η^1-like Coordination	η^2-like Coordination	η^3-like Coordination
NaPb Type	16	8	4	4
$Na_4Sn_4 \cdot 13NH_3$	4	/	/	4
$Na_4Sn_4 \cdot 11.5NH_3$	5	/	2	3
$Cs_4Pb_4 \cdot 5NH_3$	12	6	5	1
$Rb_4Sn_4 \cdot 2NH_3/Cs_4Sn_4 \cdot 2NH_3/Rb_4Pb_4 \cdot 2NH_3$	14	7	4	3

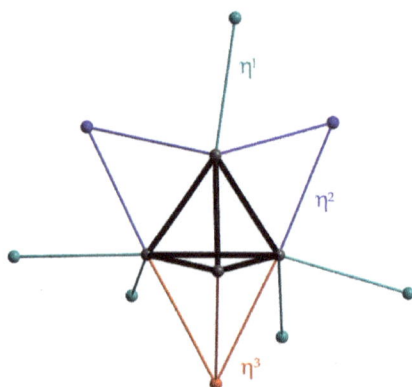

Figure 8. Different coordination modes of cations shown on the example of NaPb.

Table 5. Coordination number of the cations in NaPb and related ammoniates.

Compound	CN_{total} of Cations	A^+-E^- Contacts	A^+-NH_3 Contacts
NaPb Type	6–8	6–8	0
$Na_4Sn_4 \cdot 13NH_3$	7	3	4
$Na_4Sn_4 \cdot 11.5NH_3$	5–6	2–3	0–3
$Cs_4Pb_4 \cdot 5NH_3$	9–10	4–5	4–6
$Rb_4Sn_4 \cdot 2NH_3 / Cs_4Sn_4 \cdot 2NH_3 / Rb_4Pb_4 \cdot 2NH_3$	8–11	5–7	2–4

4.2. KGe Type Analogies

Binary alkali metal compounds of silicon and germanium with the nominal composition AB (A = K–Cs) crystallize in the KGe structure type (Figure 5, for the corresponding literature see Table 1) [11–17]. Table 6 shows the number of cations coordinated to the $[E_4]^{4-}$ cages. Here again, the decrease of the CN is directly related to the content of ammonia in the solvate structure. Like in the NaPb structure type, four cations coordinate η^3-like to all triangular faces of the cages. However, unlike the NaPb type, no η^2-like bonded cations are present in this solid state structure. Here, only single cation-anion contacts between the vertex atoms and the cations are built. The CN sums up to 16. The cationic environment of the $[E_4]^{4-}$ clusters of the compounds $Cs_4Ge_4 \cdot 9NH_3$ and $[Li(NH_3)_4]_4Sn_4 \cdot 4NH_3$ is very similar to those of the KGe structure. It consists of four η^3-like bonded cations/cationic complexes that are situated on the faces of the cages. However, in these ammoniates, every vertex is only coordinated by one cation/cationic complex instead of three. Thus, the CN has a value of eight for the anionic clusters in $Cs_4Ge_4 \cdot 9NH_3$ and $[Li(NH_3)_4]_4Sn_4 \cdot 4NH_3$ (Table 6). The situation for the $[Si_4]^{4-}$ tetrahedra in $Cs_4Si_4 \cdot 7NH_3$ looks a bit different. Here also four cations which span the triangular faces of the cages are found next to three (for Si_4 (**1**)) and four (for Si_4 (**2**)) η^1-like bonded cations, respectively (Figure 5a,b). The coordination spheres of the cages are completed by two and three cations, respectively, which coordinate to edges (η^2-like) of the cages. This kind of coordination is more prevalent in the NaPb structure type (Figure 2). The CN of the $[Si_4]^{4-}$ cages sums up to 9/11 (Table 6).

Cs ammoniates of all group 14 elements are now known ($Cs_4Si_4 \cdot 7NH_3$, $Cs_4Ge_4 \cdot 9NH_3$, $Cs_4Sn_4 \cdot 2NH_3$ and $Cs_4Pb_4 \cdot 5NH_3$), which allows for comparison of the coordination number of the Cs cation. As mentioned in the introduction, the $[E_4]^{4-}$ cages can be considered as roughly spherical with a radius calculated from the distance of the center of the tetrahedron to the edges (averaged distances) plus the *van der Waals* radius of the particular element [25]. Naturally, the sizes of the silicide (radius r: 3.58 Å) and the germanide (r: 3.67 Å) clusters are smaller than those of the stannide (r: 3.96 Å) and plumbide (r: 3.90 Å) clusters. This affects the CN of the cation. As listed in Tables 5 and 7, which show the coordination number of the cations in NaPb/KGe and the related ammoniates, the CN of Cs^+ amounts to 9–11 in $Cs_4Pb_4 \cdot 5NH_3$ and $Cs_4Sn_4 \cdot 2NH_3$. In $Cs_4Si_4 \cdot 7NH_3$ and $Cs_4Ge_4 \cdot 9NH_3$ the CN sums up to 10–13 and thus is significantly higher. In addition, the total coordination number consists of the cation-anion (A^+-E^-) and the cation-nitrogen contacts (A^+-NH_3). Considering the A^+-E^- contacts in $Cs_4Si_4 \cdot 7NH_3$ and $Cs_4Ge_4 \cdot 9NH_3$, remarkably fewer (2–7) can be found in comparison to the ion-dipole interactions (4–9) between the cesium cations and the ammonia molecules of crystallization. In contrast, in $Cs_4Pb_4 \cdot 5NH_3$ and $Cs_4Sn_4 \cdot 2NH_3$ more A^+-E^- contacts (4–7) than A^+-NH_3 (2–6) interactions occur. The reduced cation-NH_3 contacts in the solvate structures of the heavier homologues tin and lead indicates that the size of the clusters has a significant impact on the quantity of ammonia molecules of crystallization that coordinate to the cesium cation and thus complete the coordination sphere.

Altogether, it is shown that the content of ammonia molecules of crystallization directly correlates with the CN of the cages to cations (see Section 4.1). This means that the presence of ammonia molecules results in broken anion-cation contacts within the ionic framework.

Table 6. Coordination number of the $[E_4]^{4-}$ cages in KGe and related ammoniates.

Compound	Coordination Number (CN) E^--A^+	η^1-like Coordination	η^2-like Coordination	η^3-like Coordination
KGe Type	16	12	/	4
$Cs_4Si_4 \cdot 7NH_3$	9/11	3/4	2/3	4/4
$Cs_4Ge_4 \cdot 9NH_3$	8	4	/	4
$[Li(NH_3)_4]_4Sn_4 \cdot 4NH_3$	8	4	/	4

Table 7. Coordination number of the cations in KGe and related ammoniates.

Compound	CN_{total} of Cations	A^+-E^- Contacts	A^+-NH_3 Contacts
KGe Type	6	6	0
$Cs_4Si_4 \cdot 7NH_3$	10–13	2–7	4–9
$Cs_4Ge_4 \cdot 9NH_3$	13	4	9
$[Li(NH_3)_4]_4Sn_4 \cdot 4NH_3$	7	3	4

4.3. Effect of Additional Anions within Solvate Structures

In $K_{4.5}Sn_4(OH)_{0.5} \cdot 1.75NH_3$, the cationic environment slightly differs from the binary system NaPb and from the other solvate structures due to the presence of another anionic component, the hydroxide anion. As already mentioned in Section 3.6, the asymmetric unit consists of four crystallographically independent $[Sn_4]^{4-}$ clusters. The coordination number of three of them has a value of 14, the CN of the fourth cluster is 16. Although ammonia molecules of crystallization are present in the structure, the CN of the clusters are very similar to the CN of the binary solid state system or are rather insignificantly smaller. Mainly, the differences lie in the manner of the coordination of the cations to the cages. In $K_{4.5}Sn_4(OH)_{0.5} \cdot 1.75NH_3$, only one to two (for Sn_4 (2)) cations span triangular faces of the cage, compared to the binary phase and the other solvate structures described in Section 4.1, where four and three cations coordinate in a η^3-like fashion. The number of the η^2-like bonded cations is somewhat higher. They can be found five or rather six times. The remaining cationic environment of the stannide clusters is built up by six to nine cations, which are η^1-like attached to vertex tin atoms. In the binary system, every vertex atom of the cluster shows two single cation-anion contacts, so eight η^1-like bonded cations appear here. Altogether, the presence of another anionic component and ammonia molecules of crystallization lead to different cationic coordinations of the anionic cages compared to the other solvate structures. But taking the slight content of ammonia molecules and hydroxide anion per stannide cluster into account, it is not surprising that the number of coordinating cations is almost equal to that in the binary phase NaPb.

5. Conclusions

We investigated the relations of ammoniate crystal structures of tetrelide tetrahedranides and the corresponding binary intermetallic phases. The involvement of ammonia strongly influences the structures of the compounds due to its character of rather acting as ligand towards the alkali metal cations than as an innocent solvent molecule. This is reflected in the CN of the cations as well as the anions. For the small alkali metal cations of lithium and sodium (hard acids) this even results in a formal enlargement of the cation radius which finally ends up in the structural similarities especially for Li-ammonia containing compounds to the binaries of the heavier homologues. Additional charged anions within the solvate crystal structures influence the overall crystal structure and this leads to a different cationic coordination of the anionic cages compared to the other solvate structures.

Author Contributions: C.L. and S.G. carried out experimental work (synthesis, crystallization, X-ray structure determination), C.L. and S.G. prepared the manuscript, N.K. designed and conceived the study.

Funding: This research received no external funding.

Conflicts of Interest: The authors declare no conflict of interest.

Appendix A

Appendix A.1 Experimental Details

All operations were carried out under argon atmosphere using standard Schlenk and Glovebox techniques. Liquid ammonia was dried and stored on sodium in a dry ice cooled Dewar vessel for at least 48 h. Silicon (powder, 99%, 2N+, ABCR) and Lithium (99%, Chemmetal, Langelsheim) was used without further purification. Sodium (>98%, Merck, Deutschland) and potassium (>98%, Merck, Deutschland) were purified by liquating. Rubidium and cesium were synthesized according to Hackspill [51] and distilled for purification. [18]crown-6 was sublimated under dynamic vacuum at 353 K. [2.2.2]cryptand (ABCR) was used without further purification. In the reaction mixtures containing the two chelating agents, crystals of the composition $C_{12}H_{24}O_6 \cdot 2NH_3$ [52] and $C_{18}H_{36}O_6N_2 \cdot 2NH_3$ [52] could additionally be observed. For the reaction mixtures with ^tBuOH, surprisingly no crystal structures containing ^tBuOH or $^tBuO^-$ could be found.

Appendix A.1.1 Direct Reduction

$[Li(NH_3)_4]_4Sn_4 \cdot 4NH_3$, $Na_4Sn_4 \cdot 11.5NH_3$, $K_{4.5}Sn_4(OH)_{0.5} \cdot 1.75NH_3$ and $Cs_4Pb_4 \cdot 5NH_3$: Tin and lead, respectively, as well as the stoichiometric amount of alkali metals, were placed in a three times baked out Schlenk vessel in a glovebox under argon atmosphere. For the synthesis of $[Li(NH_3)_4]_4Sn_4 \cdot 4NH_3$, $Na_4Sn_4 \cdot 11.5NH_3$ and $K_{4.5}Sn_4(OH)_{0.5} \cdot 1.75NH_3$ ^tBuOH was additionally placed in the Schlenk vessel (the difference in applied amount of alkali metal and crystallized stoichiometry is explainable due to traces of water in ^tBuOH despite intensive drying). About 10 mL of dry liquid ammonia was condensed on the mixture at 195 K. The appropriate amount of ^tBuOH was added by a syringe at 195 K under immediate freezing. The blue ammonia alkali metal solution was allowed to react with ^tBuOH and tin at 236 K. Gassing was observed first and the color of the solution changed from blue to dark red within few days. After storage at 236 K for a few weeks crystals of the above discussed compounds could be obtained.

$[Li(NH_3)_4]_4Sn_4 \cdot 4NH_3$: 0.3 g Sn (2.6 mmol), 0.0905 g Li (13.1 mmol) and 1 mL ^tBuOH (0.77 g, 10.4 mmol).

$Na_4Sn_4 \cdot 11.5NH_3$: 0.95 g Sn (8.0 mmol), 0.4 g Na (17.4 mmol) and 1 mL ^tBuOH (0.77 g, 10.4 mmol). For N6, a split position was introduced as well as a SIMU restraint.

$K_{4.5}Sn_4(OH)_{0.5} \cdot 1.75NH_3$: 0.475 g Sn (4.0 mmol), 0.340 g K (8.7 mmol) and 0.5 mL ^tBuOH (0.385 g, 5.2 mmol).

$Cs_4Pb_4 \cdot 5NH_3$: 0.1729 g Pb (0.835 mmol), 0.1109 g Cs (0.835 mmol).

Appendix A.1.2 Solvolysis

Synthesis of the precursor $Cs_{12}Si_{17}$ and $Cs_{12}Ge_{17}$: For $Cs_{12}Si_{17}$, Cs (1.539 g, 11.581 mmol) and Si (0.461 g, 16.407 mmol), for $Cs_{12}Ge_{17}$, Cs (1.127 g, 8.481 mmol) and Ge (0.873 g, 12.015 mmol) were enclosed in tantalum containers and jacketed in an evacuated ampoule of fused silica. The containers were heated to 1223 K at a rate of 25 K·h^{-1}. The temperature was kept for 2 h. The ampoule was cooled down with a rate of 20 K·h^{-1}. The precursors were stored in a glove box under argon.

$Cs_4Si_4·7NH_3$ and $Cs_4Ge_4·9NH_3$: 50 mg of each precursor were dissolved in about 15 ml of liquid ammonia together with two chelating agents, [18]crown-6/dicyclohexano[18]crown-6 and [2.2.2]cryptand. The rufous solutions were stored at 197 K. After several months, very few crystals of the above discussed compounds could be obtained.

$Cs_4Si_4·7NH_3$: 50 mg (0.0245 mmol) of $Cs_{12}Si_{17}$, 9.4 mg (0.0252 mmol) dicyclohexano[18]crown-6 and 18.5 mg (0.0491 mmol) [2.2.2]cryptand. For Cs5 a split position was introduced and a SIMU restraint was applied.

$Cs_4Ge_4·9NH_3$: 50 mg (0.025 mmol) of $Cs_{12}Ge_{17}$, 0.0513 mg (0.194 mmol) [18]crown-6 and 0.044 mg (0.116 mmol) [2.2.2]cryptand. To prevent N2 and N3 to go non-positive definite (N.P.D.), the two atoms were refined isotropically and the atom radii were fixed at 0.05.

References

1. Korber, N. Metal Anions: Defining the Zintl Border. *Z. Anorg. Allg. Chem.* **2012**, *638*, 1057–1060. [CrossRef]
2. Nesper, R. The Zintl-Klemm Concept-A Historical Survey. *Z. Anorg. Allg. Chem.* **2014**, *640*, 2639–2648. [CrossRef]
3. Dubois, J.M.E.; Belin-Ferre, E.E. *Complex Metallic Alloys: Fundamentals and Applications*; Wiley-VCH Verlag GmbH: Weinheim, Germany, 2011.
4. Wesbrook, J.H.; Fleischer, R.L. *Intermetallic Compounds, Principles and Practice*; Wiley: New York, NY, USA, 2002; Volume 3.
5. Guloy, A.M. Polar Intermetallics and Zintl Phases along the Zintl Border. In *Inorganic Chemistry in Focus III*; Wiley-VCH Verlag GmbH & Co. KGaA: Weinheim, Germany, 2006.
6. Zintl, E. Intermetallische Verbindungen. *Angew. Chem.* **1939**, *52*, 1–6. [CrossRef]
7. Klemm, W. Metalloids and their compounds with the alkali metal. *Proc. Chem. Soc. Lond.* **1958**, *12*, 329–341.
8. Gärtner, S.; Korber, N.; Poeppelmeier, K.E.; Reedijk, J.E. *Main-Group Elements, Comprehensive Inorganic Chemistry II*, 2nd ed.; Elsevier Ltd.: Amsterdam, The Netherlands, 2013; Volume 140.
9. Fässler, T.F. *Structure and Bonding*; Springer: Berlin/Heidelberg, Germany, 2011; Volume 140.
10. Marsh, R.E.; Shoemaker, D.P. The crystal structure of NaPb. *Acta Cryst.* **1953**, *6*, 197–205. [CrossRef]
11. Busmann, E. Das Verhalten der Alkalimetalle zu Halbmetallen. X. Die Kristallstrukturen von KSi, RbSi, CsSi, KGe, RbGe und CsGe. *Z. Anorg. Allg. Chem.* **1961**, *313*, 90–106. [CrossRef]
12. Von Schnering, H.G.; Schwarz, M.; Chang, J.-H.; Peters, K.; Peters, E.-M.; Nesper, R. Refinement of the crystal structures of the tetrahedrotetrasilicides K_4Si_4, Rb_4Si_4 and Cs_4Si_4. *Z. Kristallogr. NCS* **2005**, *220*, 525–527. [CrossRef]
13. Von Schnering, H.G.; Llanos, J.; Chang, J.H.; Peters, K.; Peters, E.M.; Nesper, R. Refinement of the crystal structures of the tetrahedro-tetragermanides K_4Ge_4, Rb_4Ge_4 and Cs_4Ge_4. *Z. Kristallogr. NCS* **2005**, *220*, 324–326. [CrossRef]
14. Schäfer, R.; Klemm, W. Das Verhalten der Alkalimetalle zu Halbmetallen. IX. Weitere Beiträge zur Kenntnis der Silicide und Germanide der Alkalimetalle. *Z. Anorg. Allg. Chem.* **1961**, *312*, 214–220. [CrossRef]
15. Hohmann, E. Silicide und Germanide der Alkalimetalle. *Z. Anorg. Allg. Chem.* **1948**, *257*, 113–126. [CrossRef]
16. Witte, J.; von Schnering, H.G. Die Kristallstruktur von NaSi und NaGe. *Z. Anorg. Allg. Chem.* **1964**, *327*, 260. [CrossRef]
17. Goebel, T.; Ormeci, A.; Pecher, O.; Haarmann, F. The Silicides M_4Si_4 with M = Na, K, Rb, Cs, and Ba_2Si_4–NMR Spectroscopy and Quantum Mechanical Calculations. *Z. Anorg. Allg. Chem.* **2012**, *638*, 1437–1445. [CrossRef]
18. Baitinger, M.; Grin, Y.; von Schnering, H.G.; Kniep, R. Crystal structure Rb_4Sn_4 and Cs_4Sn_4. *Z. Kristallogr. New Cryst. Struct.* **1999**, *214*, 457–458.

19. Grin, Y.; Baitinger, M.; Kniep, R.; von Schnering, H.G. Redetermination of the crystal structure of tetrasodium tetrahedrotetrastannide, Na_4Sn_4 and tetrapotassium tetrahedro-tetrastannide, K_4Sn_4. *Z. Kristallogr New Cryst. Struct.* **1999**, *214*, 453–454. [CrossRef]

20. Baitinger, M.; Peters, K.; Somer, M.; Carrillo-Cabrera, W.; Grin, Y.; Kniep, R.; von Schnering, H.G. Crystal structure of tetrarubidium tetrahedro-tetraplumbide, Rb_4Pb_4 and of tetracaesium tetrahedro-telraplumbide, Cs_4Pb_4. *Z. Kristallogr. NCS* **1999**, *214*, 455–456.

21. Hewaidy, I.F.; Busmann, E.; Klemm, W. Die Struktur der AB-Verbindungen der schweren Alkalimetalle mit Zinn und Blei. *Z. Anorg. Allg. Chem.* **1964**, *328*, 283–293. [CrossRef]

22. Goebel, T.; Prots, Y.; Haarmann, F. Refinement of the crystal structure of tetrasodium tetrasilicide, Na_4Si_4. *Z. Kristallogr. NCS* **2008**, *223*, 187–188. [CrossRef]

23. Evers, J.; Oehlinger, G.; Sextl, G. LiSi, a Unique Zintl Phase—Although Stable, It Long Evaded Synthesis. *Eur. J. Solid State Inorg. Chem.* **1997**, *34*, 773–784.

24. Menges, E.; Hopf, V.; Schaefer, H.; Weiss, A. Die Kristallstruktur von LiGe—ein neuartiger, dreidimensionaler Verband von Element (IV)-atomen. *Z. Naturf. B* **1969**, *24*, 1351–1352. [CrossRef]

25. Holleman, A.F.; Wiberg, E.; Wiberg, N. *Anorganische Chemie*, 103rd ed.; Walter de Gruyter GmbH: Berlin, Germany, 2017.

26. Müller, W.; Schäfer, H. Crystal-structure of LiSn. *Z. Naturforsch. B* **1973**, *B 28*, 246–248.

27. Müller, W.; Volk, K. Die Struktur des beta-NaSn. *Z. Naturforsch.* **1977**, *B 32*, 709–710.

28. Nowotny, H. The structure of LiPb. *Z. Metallkunde* **1941**, *33*, 388.

29. Neumeier, M.; Fendt, F.; Gaertner, S.; Koch, C.; Gaertner, T.; Korber, N.; Gschwind, R.M. Detection of the Elusive Highly Charged Zintl Ions $[Si_4]^{4-}$ and $[Sn_4]^{4-}$ in Liquid Ammonia by NMR Spectroscopy. *Angew. Chem. Int. Ed.* **2013**, *52*, 4483–4486. [CrossRef] [PubMed]

30. Wiesler, K.; Brandl, K.; Fleischmann, A.; Korber, N. Tetrahedral $[Tt_4]^{4-}$ Zintl Anions through Solution Chemistry: Syntheses and Crystal Structures of the Ammoniates $Rb_4Sn_4\cdot2NH_3$, $Cs_4Sn_4\cdot2NH_3$, and $Rb_4Pb_4\cdot2NH_3$. *Z. Anorg. Allg. Chem.* **2009**, *635*, 508–512. [CrossRef]

31. Fleischmann, A. Synthese und Strukturelle Charakterisierung homoatomarer Polyanionen der vierten und fünften Hauptgruppe durch Reduktion in flüssigem Ammoniak. Ph.D. Thesis, University of Regensburg, Regensburg, Germany, 2002.

32. Waibel, M.; Fässler, T.F. First Incorporation of the Tetrahedral $[Sn_4]^{4-}$ Cluster into a Discrete Solvate $Na_4[Sn_4](NH_3)_{13}$ from Solutions of Na_4Sn_4 in Liquid Ammonia. *Z. Naturforsch. B* **2013**, *68*, 732–734. [CrossRef]

33. Lorenz, C.; Gärtner, S.; Korber, N. $[Si_4]^{4-}$ in Solution–First Solvate Crystal Structure of the Ligand-free Tetrasilicide Tetraanion in $Rb_{1.2}K_{2.8}Si_4\cdot7NH_3$. *Z. Anorg. Allg. Chem.* **2017**, *643*, 141–145. [CrossRef]

34. Fendt, F. Untersuchungen zum Lösungs-und Reaktionsverhalten von Polystanniden und-siliciden in flüssigem Ammoniak. Ph.D. Thesis, University of Regensburg, Regensburg, Germany, 2016.

35. Joseph, S.; Suchentrunk, C.; Kraus, F.; Korber, N. $[Si_9]^{4-}$ Anions in Solution–Structures of the Solvates $Rb_4Si_9\cdot4.75 NH_3$ and [Rb (18-crown-6)] $Rb_3Si_9\cdot4NH_3$, and Chemical Bonding in $[Si_9]^{4-}$. *Eur. J. Inorg. Chem.* **2009**, *2009*, 4641–4647. [CrossRef]

36. Joseph, S.; Suchentrunk, C.; Korber, N. Dissolving Silicides: Syntheses and Crystal Structures of New Ammoniates Containing Si_5^{2-} and Si_9^{4-} Polyanions and the Role of Ammonia of Crystallisation. *Z. Naturforsch.* **2010**, *B65*, 1059–1065. [CrossRef]

37. Suchentrunk, C.; Daniels, J.; Somer, M.; Carrillo-Cabrera, W.; Korber, N. Synthesis and Crystal Structures of the Polygermanide Ammoniates $K_4Ge_9\cdot9 NH_3$, $Rb_4Ge_9\cdot5 NH_3$ And $Cs_6Ge_{18}\cdot4 NH_3$. *Z. Naturforsch.* **2005**, *B60*, 277–283.

38. Korber, N.; Fleischmann, A. Synthesis and crystal structure of $[Li(NH_3)_4]_4[Sn_9]\cdot NH_3$ and $[Li(NH_3)_4]_4[Pb_9]\cdot NH_3$. *Dalton Trans.* **2001**, *4*, 383–385. [CrossRef]

39. Carrillo-Cabrera, W.; Aydemir, U.; Somer, M.; Kircali, A.; Fassler, T.F.; Hoffmann, S.D. $Cs_4Ge_9\cdot$en: A Novel Compound with $[Ge_9]^{4-}$ Clusters–Synthesis, Crystal Structure and Vibrational Spectra. *Z. Anorg. Allg. Chem.* **2007**, *633*, 1575–1580. [CrossRef]

40. Somer, M.; Carrillo-Cabrera, W.; Peters, E.M.; Peters, K.; von Schnering, H.G. Tetrarubidium Nonagermanide (4−) Ethylenediamine, $Rb_4[Ge_9][en]$. *Z. Anorg. Allg. Chem.* **1998**, *624*, 1915–1921. [CrossRef]

41. Edwards, P.A.; Corbett, J.D. Stable homopolyatomic anions. Synthesis and crystal structures of salts containing the pentaplumbide (2-) and pentastannide (2-) anions. *Inorg. Chem.* **1977**, *16*, 903–907. [CrossRef]

42. Goicoechea, J.M.; Sevov, S.C. Naked Deltahedral Silicon Clusters in Solution: Synthesis and Characterization of $Si_9{}^{3-}$ and $Si_5{}^{2-}$. *J. Am. Chem. Soc.* **2004**, *126*, 6860–6861. [CrossRef] [PubMed]

43. Suchentrunk, C.; Korber, N. $Ge_5{}^{2-}$ Zintl anions: synthesis and crystal structures of [K([2.2.2]-crypt)]$_2$ $Ge_5\cdot4NH_3$ and [Rb([2.2.2]-crypt)]$_2Ge_5\cdot4NH_3$. *New J. Chem.* **2006**, *30*, 1737–1739. [CrossRef]

44. Friedrich, U.; Korber, N. $Cs_5Sn_9(OH)\cdot4NH_3$. *Acta Crystallogr. E* **2014**, *70*, i29. [CrossRef] [PubMed]

45. Kottke, T.; Stalke, D. Crystal handling at low temperatures. *J. Appl. Crystallogr.* **1993**, *26*, 615–619. [CrossRef]

46. Stalke, D. Cryo crystal structure determination and application to intermediates. *Chem. Soc. Rev.* **1998**, *27*, 171–178. [CrossRef]

47. Rigaku. *Crysalis pro, Version 1.171.38.46*; Agilent Technologies: Santa Clara, CA, USA, 2017.

48. X-RED. *STOE & Cie GmbH Darmstadt 1998*; X-RED Data Reduction for STADI4 and IPDS: Darmstadt, Germany, 1998.

49. Fendt, F.; Koch, C.; Gartner, S.; Korber, N. Reaction of $Sn_4{}^{4-}$ in liquid ammonia: the formation of $Rb_6[(\eta^2\text{-}Sn_4)$ $Zn\,(\eta^3\text{-}Sn_4)]\cdot5NH_3$. *Dalton Trans.* **2013**, *42*, 15548–15550. [CrossRef] [PubMed]

50. Pearson, R.G. Hard and Soft Acids and Bases. *J. Am. Chem. Soc.* **1963**, *85*, 3533–3539. [CrossRef]

51. Hackspill, L. Some properties of the alkali metals. *Helv. Chim. Acta* **1928**, *11*, 1.

52. Suchentrunk, C.; Rossmier, T.; Korber, N. Crystal structures of the [18]-crown-6 ammoniate $C_{12}H_{24}O_6\cdot2NH_3$ and the cryptand [2.2.2] ammoniate $C_{18}H_{36}O_6N_2\cdot2NH_3$. *Z. Kristallogr.* **2006**, *221*, 162–165.

Communication

Ba4[Mn3N6], a Quasi-One-Dimensional Mixed-Valent Nitridomanganate (II, IV)

Alexander Ovchinnikov [1,2], Matej Bobnar [1], Yurii Prots [1], Walter Schnelle [1], Peter Höhn [1,*] and Yuri Grin [1]

[1] Max-Planck-Institut für Chemische Physik fester Stoffe, Nöthnitzer Straße 40, 01187 Dresden, Germany; alexovc@udel.edu (A.O.); bobnar@cpfs.mpg.de (M.B.); prots@cpfs.mpg.de (Y.P.); schnelle@cpfs.mpg.de (W.S.); grin@cpfs.mpg.de (Y.G.)

[2] Department of Chemistry and Biochemistry, University of Delaware, Newark, DE 19716, USA

* Correspondence: hoehn@cpfs.mpg.de; Tel.: +49-351-4646-2229

Received: 25 April 2018; Accepted: 18 May 2018; Published: 25 May 2018

Abstract: The mixed-valent nitridomanganate $Ba_4[Mn_3N_6]$ was prepared using a gas–solid high temperature route. The crystal structure was determined employing high resolution synchrotron powder diffraction data: space group $Pbcn$, $a = 9.9930(1)$ Å, $b = 6.17126(8)$ Å, $c = 14.4692(2)$ Å, $V = 892.31(2)$ Å^3, $Z = 4$. The manganese atoms in the structure of $Ba_4[Mn_3N_6]$ are four-fold coordinated by nitrogen forming infinite corrugated chains of edge-sharing $[MnN_4]$ tetrahedra. The chains demonstrate a complete charge order of Mn species. Magnetization measurements and first principle calculations indicate quasi-one dimensional magnetic behavior. In addition, chemical bonding analysis revealed pronounced Mn–Mn interactions along the chains.

Keywords: nitridometalate; crystal structure; powder diffraction; magnetism

1. Introduction

Low-dimensional magnetic systems, such as spin chains, ladders, or planes, attract much attention as perspective materials for a wide range of applications, e.g., in spintronics, quantum computing, and information storage technologies [1,2]. Such quantum magnets may display exotic physical phenomena including spin liquid behavior [3], spin-orbital Mott insulating state [4], and topological excitations [5]. Since the decrease of dimensionality implies a spatial spin confinement, the role of fluctuations becomes significant in these systems. Such fluctuations are spin-dependent and most important for $S = \frac{1}{2}$ and $S = 1$ systems. Therefore, the electronic state of the constituting magnetic atoms, along with the magnetic topology, determines the behavior of a particular system.

Low-dimensional quantum magnets have been mainly explored in the families of halides [6], oxides [7], and higher chalcogenides [8], but little is known about the realization of such systems in nitrides. Since multicomponent nitrides often demonstrate low-dimensional crystallographic arrangements of transition-metal atoms along with low coordination numbers and oxidation states of the latter [9–11], they represent a natural platform to probe low-dimensional magnetism. However, the preparation of single-phase nitrides and their inherent instability make the study of this class of materials highly challenging.

In this contribution, we report on the synthesis and characterization of the first chain alkaline-earth nitridomanganate with a quasi-one-dimensional magnetic behavior.

2. Materials and Methods

Synthesis of Ba_2N. All manipulations except the high-temperature treatment were done inside an Ar-filled glovebox due to the high air- and moisture-sensitivity of most of the materials. Barium nitride, Ba_2N, was prepared by annealing Ba lumps (99.9%, Alfa Aesar, Thermo Fisher (Kandel) GmbH, Karlsruhe, Germany) under N_2 stream (Praxair Deutschland GmbH, Dresden, Germany, 99.9999%, additionally purified by molecular sieves and a BTS-catalyst) at 973 K for 12 h, followed by cooling down to room temperature under Ar. The resulting black soft powder was single-phase according to powder X-ray diffraction (PXRD).

Synthesis of $Ba_4[Mn_3N_6]$. Ba_2N (Figure S1) and Mn powder (Alfa, 99.9998%) were mixed in the ratio Ba:Mn = 4.04:3 in an agate mortar and thoroughly ground. The excess of Ba_2N was employed to compensate for evaporation at high temperatures. The mixture was pelletized and annealed in a Ta crucible under a constant N_2 flow (7 mL/min) at T = 1023–1123 K for 108 h in total, with several intermediate re-grindings. The resulting sample was almost single-phase. The intensity of the strongest impurity peak was lower than 3% of the most intense peak of the main phase. The impurity reflections could be easily distinguished from those of the main phase by tracking the evolution of the PXRD patterns upon annealing, however, they could not be assigned to any known phases. Annealing times longer than that in the above-given protocol led to gradual decomposition of the main phase and to partial amorphisation of the sample. The composition of the sample and the absence of potential impurity elements were confirmed by chemical analysis (Table S1).

Powder X-ray diffraction (PXRD). Laboratory PXRD patterns were collected on a Huber G670 imaging plate Guinier camera ($CuK_{\alpha1}$ radiation, Huber Diffraktionstechnik GmbH & Co. KG, Rimsting, Germany). Powder samples were enclosed between two Kapton foils sealed with vacuum grease to reduce contact with air. Synchrotron PXRD data were collected at the ID22 beamline of the European Synchrotron Radiation Facility (ESRF, Grenoble, France). Samples were sieved to a particle size of less than 50 μm and enclosed in glass capillaries (d = 0.3 mm) sealed with *Picein*. Preliminary data processing was performed in the WinXPow program suite [12]. Crystal structure solution was accomplished using direct methods as implemented in EXPO2009 [13]. Rietveld refinement was performed with the Jana2006 program [14]. Further details on the crystal structure investigations can be obtained from the Fachinformationszentrum Karlsruhe, 76344 Eggenstein–Leopoldshafen, Germany (fax: (+49)7247-808-666; email: crysdata@fiz-karlsruhe.de, http://www.fiz-karlsruhe.de/request_for_deposited_data.html) on quoting the depository number CSD-434473.

Chemical analysis. Chemical analysis was performed for the constituting elements (Ba, Mn, N), as well as for expected impurities (C, H, O, Ta from the crucible). Non-metals were analyzed by a carrier-gas hot-extraction technique on LECO TCH 600 (N, H, O, LECO Corporation, Saint Joseph, MI, USA) and LECO C200 (C, LECO Corporation, Saint Joseph, MI, USA) analyzers. The metal content was determined by inductively coupled plasma optical emission spectroscopy (ICP-OES) on an Agilent Technologies 5100 spectrometer (Agilent technologies, Santa Clara, CA, USA).

Differential thermal analysis and thermogravimetry (DTA-TG). Thermal behavior was studied by means of DTA/TG measurements on a Netzsch STA 449C calorimetric setup (NETZSCH-Gerätebau GmbH, Selb, Germany) in loosely closed Ta crucibles under dynamic Ar atmosphere. To prevent sample degradation, the measurements were done inside an Ar-filled glovebox.

Electrical resistivity measurements. Electrical resistivity was measured on a cold-pressed pellet in a sapphire die cell within a cryostat using a four-contact Van-der-Pauw method. The setup was mounted inside an Ar-filled glovebox. The sample was thoroughly ground and sieved before the measurements. Only the fraction with the particle size between 20 μm and 50 μm was used in order to achieve a higher packing density and reduce the grain boundary effects.

Magnetization measurements. Temperature dependence of magnetic susceptibility was measured on a powder sample enclosed in a sealed pre-calibrated quartz tube under 400 mbar of He on a SQUID magnetometer (MPMS-XL7, Quantum Design Inc., San Diego, CA, USA) in external fields between 10 mT and 7 T within the temperature range 1.8–400 K. High-temperature magnetization measurements were performed in the temperature range 320–575 K. All data were corrected for the container diamagnetism. The Honda–Owen correction ("extrapolation to a large field") was applied to take the possible contributions of ferro- or ferrimagnetic impurities into account [15].

Computational details. Spin-polarized electronic structure calculations were performed at the scalar relativistic level within the L(S)DA approach employing the FPLO-9 [16] or the TB-LMTO-ASA [17] code. The PW92 parametrization [18] of the LSDA functional was used in FPLO and the von Barth-Hedin parametrization [19] was employed in LMTO. Blöchl corrected linear tetrahedron method with a $8 \times 12 \times 6$ k-mesh was employed after checking for convergence with respect to the number of k-points. For the LMTO calculations, experimentally obtained lattice parameters and atomic coordinates were used. The radial scalar-relativistic Dirac equation was solved to get the partial waves. The calculation within the atomic sphere approximation (ASA) includes corrections for the neglect of interstitial regions and partial waves of higher order [20], hence an addition of empty spheres in the case of $Ba_4[Mn_3N_6]$ was not necessary. The following radii of atomic spheres were applied for the calculations on: r(Ba1) = 2.116 Å; r(Ba21) = 2.187 Å, r(Mn1) = 1.415 Å, r(Mn2) = 1.217 Å, r(N1) = 1.094 Å, r(N2) = 1.003 Å and r(N3) = 1.044 Å. A basis set containing Ba($6s,5d$), Mn($4s, 4p, 3d$) and N($2s,2p$) states was employed for the self-consistent calculations with the Ba($6p,4f$) and N($3d$) functions being downfolded. The electronic structures calculated with the two codes were found to be consistent. For the analysis of the Mn–Mn interactions, Crystal Orbital Hamilton Population (COHP) [21] analysis was performed using the built-in procedure in the TB-LMTO-ASA program. The topology of electron density was analyzed with the program Dgrid [22]. The calculated electron density was integrated in basins, bounded by zero-flux surfaces in the density gradient field [23]. This technique provides electron counts for each atomic basin revealing the effective charges of the QTAIM atoms.

3. Results and Discussion

Crystal structure determination. $Ba_4[Mn_3N_6]$ was obtained as an almost phase-pure microcrystalline product. Attempts to grow single crystals were not successful. Therefore, crystal structure determination was accomplished based on high resolution synchrotron PXRD data (Figure 1). The reflections of the major phase were indexed in the orthorhombic crystal system with the lattice parameters $a = 9.9930(1)$ Å, $b = 6.17126(8)$ Å, $c = 14.4692(2)$ Å. Extinction conditions were consistent only with the space group *Pbcn* (#60). Crystal structure solution using direct methods provided the positions of all metal atoms and a part of the nitrogen atoms. The remaining nitrogen positions were located in a subsequent difference Fourier synthesis (Figure 2). For Ba atoms, an anisotropic refinement was possible. Mn and N atoms were refined isotropically. In addition, atomic displacement parameters for all N atoms were constrained to be the same in the final cycles of the refinement. The crystallographic data are listed in Table 1, atomic positions in Table 2, atomic displacement parameters in Table 3, and selected bond lengths/angles in Figure 3 and Table S2, respectively.

Crystal structure description. The crystal structure of $Ba_4[Mn_3N_6]$ represents a new structure type and can be viewed as consisting of corrugated chains of edge-sharing [MnN$_4$] tetrahedra running along [001], and Ba atoms embedded in-between the chains (Figure 2a,b). A similar structural motif is observed for the nitridometalates $AE_3[M_2N_4]$ (AE = Sr, Ba, M = Al, Ga, Ge/Mg) [24–29] (Figure 2c), though the degree of the chain corrugation is weaker in these compounds. Perfectly linear chains of edge-sharing [Fe^{3+}N$_4$] tetrahedra are present in the crystal structure of the lithium nitridoferrate $Li_3[FeN_2]$ [30] (Figure 2d).

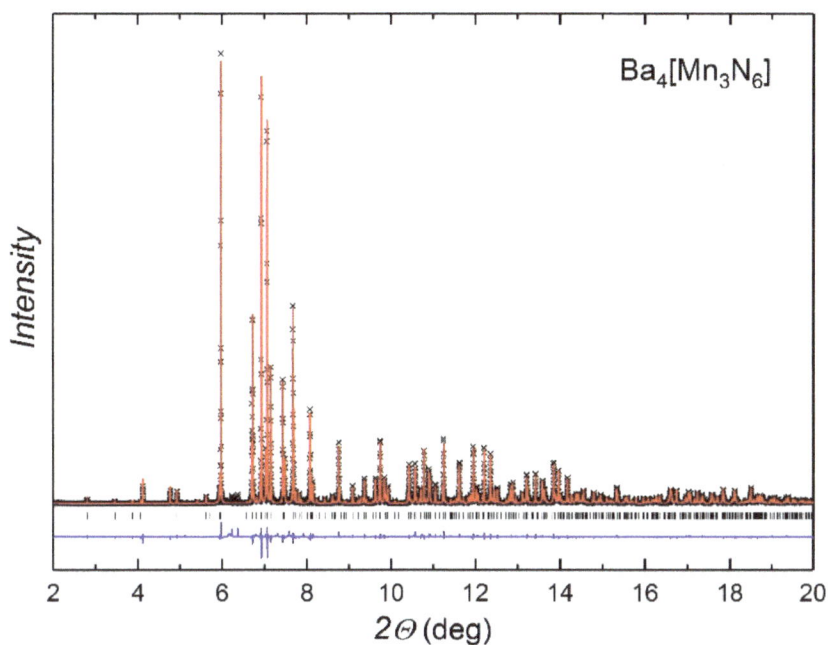

Figure 1. Synchrotron PXRD pattern of $Ba_4[Mn_3N_6]$ (λ = 0.35434 Å) with experimental points shown in black, calculated pattern after Rietveld refinement in red, and difference curve in blue. Tick marks denote the calculated positions of Bragg reflections.

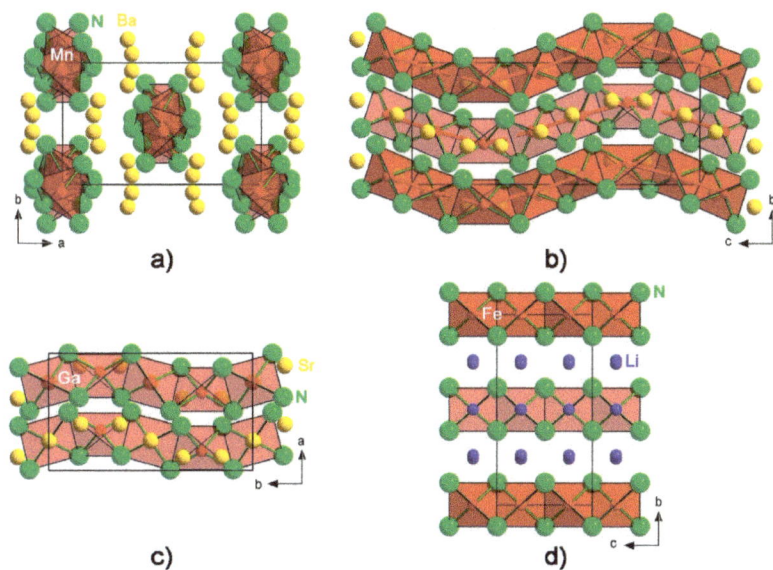

Figure 2. Crystal structure of $Ba_4[Mn_3N_6]$ viewed along (**a**) [001] and (**b**) [100], for comparison, chains of edge-sharing tetrahedra in (**c**) $Sr_3[Ga_2N_4]$ [26] and (**d**) $Li_3[FeN_2]$ [30] are drawn on the same scale.

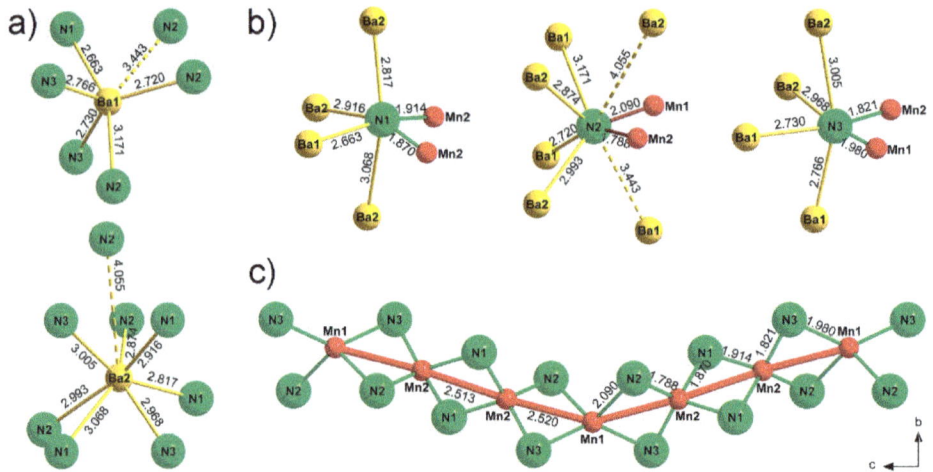

Figure 3. Coordination environment of (**a**) barium, (**b**) nitrogen, and (**c**) manganese atoms in $Ba_4[Mn_3N_6]$. Distances are given in Å.

Table 1. Crystallographic Data and Experimental Details for $Ba_4[Mn_3N_6]$ at 298 K.

Composition	$Ba_4Mn_3N_6$
Molecular weight/g mol^{-1}	798.16
Space group	*Pbcn* (#60)
Lattice parameters [1]	
a/Å	9.9930(1)
b/Å	6.17126(8)
c/Å	14.4692(2)
V/Å^3	892.31(2)
Z	4
ρ_{calcd}/g cm^{-3}	5.94
T/K	298
Device	beamline ID22, ESRF
Radiation, λ/Å	0.35434
2Θ max/°	38
2Θ step/°	0.002
R_I/R_p	0.044/0.058
Residual electron density peaks/e Å^{-3}	+1.10, −0.94

[1] The standard deviations include the Bérar–Lelann's correction [31].

Table 2. Atomic Positions and Isotropic (Equivalent) Displacement Parameters (Å^2) for $Ba_4[Mn_3N_6]$ [1].

Atom	Site	x	y	z	$U_{iso}*/U_{eq}$
Ba1	8*d*	0.82617(7)	0.68848(10)	0.19158(4)	0.00687(19)
Ba2	8*d*	0.65129(6)	0.93009(10)	0.94504(4)	0.0077(2)
Mn1	4*c*	0	0.1719(4)	1/4	0.0066(6)*
Mn2	8*d*	0.00844(13)	0.0660(3)	0.08193(10)	0.0055(4)*
N1	8*d*	0.9009(7)	0.8512(11)	0.0291(5)	0.0067(11)* [2]
N2	8*d*	0.1212(7)	0.9716(10)	0.1690(5)	0.0067* [2]
N3	8*d*	0.9288(7)	0.3008(12)	0.1347(5)	0.0067* [2]

[1] The standard deviations include the Bérar–Lelann's correction [31]; [2] $U_{iso}(N1) = U_{iso}(N2) = U_{iso}(N3)$ constrained.

Table 3. Anisotropic Displacement Parameters (Å^2) for $Ba_4[Mn_3N_6]$ [1].

Atom	U_{11}	U_{22}	U_{33}	U_{12}	U_{13}	U_{23}
Ba1	0.0088(3)	0.0061(3)	0.0058(3)	0.0003(4)	0.0016(3)	−0.0008(3)
Ba2	0.0066(3)	0.0079(3)	0.0085(4)	0.0004(3)	−0.0015(4)	−0.0011(4)

[1] The standard deviations include the Bérar-Lelann's correction [31].

Ba1 and Ba2 atoms are the (5 + 1)-fold and (7 + 1)-fold irregularly coordinated, respectively (Figure 3a). For Ba1, the five shortest Ba–N bonds range from 2.66 Å to 3.17 Å. The resulting $[Ba1N_5]$ entity resembles a distorted square pyramid, similar to the corresponding coordination polyhedron in another barium nitridomanganate, $Ba_3[MnN_3]$ [32]. The sixth nitrogen atom is located at a relatively longer distance of 3.44 Å, thereby completing the (5 + 1)-fold coordination environment. Ba2 is coordinated by seven N atoms at distances of 2.82–3.07 Å and a further one at 4.06 Å. These distances are in good agreement with the typical values found in other nitridometalates containing Barium [24,27,29,32].

All nitrogen atoms in the structure have four Ba and two Mn atoms in the closest proximity (Figure 3b). The nitrogen atoms N1 and N3 show a distorted octahedral environment, whereas the coordination environment of N2 resembles a trigonal prism. An additional Ba1 atom at 3.44 Å and the Ba2 atom at the farthest corner residing at a distance of 4.06 Å complete the (6 + 2)-fold coordination of N2. A similar coordination of nitrogen is observed in the above-mentioned $AE_3[Al_2N_4]$ (AE = Sr, Ba) compounds [24,25] and can be regarded as a strongly distorted square anti-prism. The corresponding longest distance in $Ba_3[Al_2N_4]$ is 4.04 Å [24].

Two symmetrically independent manganese atoms, Mn1 and Mn2, alternate in the sequence [–Mn1–Mn2–Mn2–] along the chain with rather short Mn–Mn distances of 2.51 and 2.52 Å (Figure 3c). These distances fall in the range of the metal-metal contacts in metallic manganese (α-Mn: 2.26–2.93 Å [33]; β-Mn: 2.36–2.68 Å [34]; $Mn_2N_{1.08}$: 2.79–2.82 Å [35]; Mn_4N: 2.74 [36]). The Mn–N contacts lie in the range 1.98–2.09 Å for Mn1 ($<d>$ = 2.04 Å) and 1.79–1.91 Å for Mn2 ($<d>$ = 1.85 Å), with the respective bond valence sums of 2.4 for Mn1 and 3.9 for Mn2 (based on the bond valence parameters from Brese and O'Keeffe [37]). Hence, the oxidation states balance based on the structural data can be expressed as $Ba^{2+}_4[Mn^{2+}N^{3-}_{4/2}][Mn^{4+}N^{3-}_{4/2}]_2$.

Distribution of oxidation states in $Ba_4[Mn_3N_6]$ is further corroborated by comparing the Mn–N bond distances with those in other nitridocompounds bearing tetrahedrally coordinated Mn atoms. The range of the Mn–N distances around Mn1 is similar to that in $Mn^{2+}[GeN_2]$ (d(Mn–N) = 2.03–2.14 Å, $<d>$ = 2.10 Å) [38] and α-$Mn^{2+}[WN_2]$ (d(Mn–N) = 2.01–2.19 Å, $<d>$ = 2.10 Å) [39]. For Mn2, no proper reference compound was found, since there are no phases known with Mn^{4+} adopting a tetrahedral environment of nitrogen ligands. However, the Mn–N bond length distribution around Mn2 resembles that of Mn atoms in $Li_7[Mn^{5+}N_4]$ (d(Mn–N) = 1.81–1.83 Å, $<d>$ = 1.82 Å) [40], with the bonds in the latter compound being shorter due to a higher oxidation state of Mn.

The significant difference of the Mn–N bond distances around the two independent Mn sites, which made the above-described oxidation state assignment possible, suggests a charge-ordering scenario, frequently observed for mixed-valent manganese oxides [41]. Thus, $K_5[Mn_3O_6]$ and $Rb_8[Mn_5O_{10}]$, possessing chains of edge-sharing $[MnO_4]$ tetrahedra, were reported to develop full charge order into di- and tri-valent manganese with the formation of the repetition units [–Mn^{3+}–Mn^{2+}–Mn^{2+}–] and [–Mn^{3+}–Mn^{2+}–Mn^{2+}–Mn^{3+}–Mn^{2+}–] for the potassium and rubidium phase, respectively [42]. In the title compound $Ba_4[Mn_3N_6]$, the repetition unit is [–Mn^{2+}–Mn^{4+}–Mn^{4+}–]. Hence, this nitridomanganate does not conform to the known tendency of transition metals to adopt lower oxidation states in nitride compounds in comparison with the oxide analogues [9].

Physical properties. The temperature dependence of the electrical resistivity for $Ba_4[Mn_3N_6]$ is shown in Figure 4. Only data above 62 K were obtained, since at lower temperatures, the resistance exceeds the maximal measurable value achievable with the employed experimental set-up.

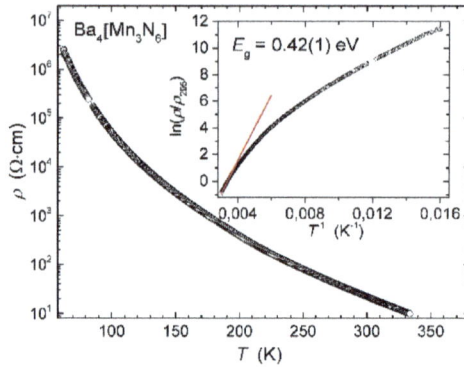

Figure 4. Temperature dependence of electrical resistivity for polycrystalline $Ba_4[Mn_3N_6]$. Inset: $\ln(\rho/\rho_{295})$ vs. T^{-1} plot (circles) with a linear fit (red line).

At room temperature, the resistivity of the sample amounts to 25 $\Omega \cdot$cm. The sample displays a distinct semiconducting behavior. As it is seen from the $\ln(\rho/\rho_{295})$ vs. T^{-1} plot, the temperature behavior of resistivity does not follow a simple Arrhenius-type dependence in a wide temperature range. Linear fitting of the high-temperature region yields an estimated bandgap of 0.42(1) eV. The plot can be linearized in the $\ln(\rho)$ vs. $T^{-1/n}$ coordinates, which is frequently discussed as an indication of variable-range hopping conduction. However, a final decision cannot be made since relatively recent calculations showed that such kind of behavior can be observed even for a traditional band transport mechanism [43]. Additional transport measurements on not yet available single crystal samples would be necessary to get a deeper insight into the electrical conduction of $Ba_4[Mn_3N_6]$.

Magnetic susceptibility of $Ba_4[Mn_3N_6]$ versus temperature is given in Figure 5. The observed weak field dependence points to a possible ferro- or ferrimagnetic impurity, most likely ferrimagnetic Mn_4N ($T_C = 740$ K) [36]. In this case, the impurity amounts to less than 0.4 mass% as can be estimated from the magnetization data. Due to the small amount, Mn_4N was not observed in the PXRD pattern of the sample under study.

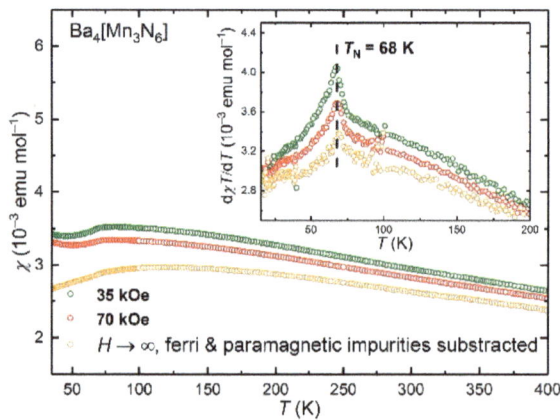

Figure 5. Temperature dependence of magnetic susceptibility for $Ba_4[Mn_3N_6]$ up to 400 K measured in different fields. Inset: $d\chi T/dT(T)$ plot emphasizing the AFM transition.

The magnetic susceptibility shows an upturn at low temperatures which is probably due to paramagnetic Mn species contained in minor secondary phase(s) or due to point defects in the main phase. After the subtraction of a Curie law with $C = 0.0211(2)$ emu mol^{-1} K, the susceptibility reveals a broad hump around 120 K and a clear decrease of $\chi(T)$ below $T_N = 68$ K, where a sharp kink is observed. Such temperature dependence is typical for low-dimensional (here a quasi-1D) magnetic systems. However, we abstain from a detailed analysis of the magnetic susceptibility data. For a deeper investigation, anisotropic magnetization data on single crystals are required.

Measurements of the susceptibility to higher temperatures are hampered by the degradation of the sample. According to the DTA/TG measurements, $Ba_4[Mn_3N_6]$ starts to lose mass at around 673 K under inert conditions (Figure S2). For these reasons, the magnetic measurements were performed up to 575 K only to avoid possible decomposition. Between 325 and 575 K, the temperature dependence of the magnetic susceptibility was found to decrease almost linearly (Figure S3).

Electronic structure and chemical bonding. Total energy calculations were performed for eight different magnetic arrangements (Figure S4) to discern the ground state. The lowest energy was found for the AFM1 structure, which displays antiferromagnetic coupling (AFM) between manganese atoms along the chains and ferromagnetic (FM) coupling between the nearest manganese sites in adjacent chains. AFM1 with FM coupled AFM chains is only 0.7 meV f.u.$^{-1}$ more stable than AFM2, possessing AFM coupled AFM chains. It is clear that the actual ground state cannot be reliably determined from LSDA calculations owing to the negligible energy difference between the two best candidates. However, the most stable solution without AFM intra-chain coupling (AFM3 in Figure S3) is by 138 meV f.u.$^{-1}$ higher in energy than AFM1. These findings emphasize the strongly one-dimensional nature of exchange interactions in $Ba_4[Mn_3N_6]$. The calculated magnetic moments and the bandgap are almost the same for AFM1 and AFM2 structures (Table 4).

Table 4. Results of the LSDA calculations for magnetic structures AFM1 and AFM2 (FPLO).

Structure	AFM1	AFM2
energy E with respect to AFM1 (meV f.u.$^{-1}$)	0	0.7
electronic bandgap E_g (eV)	0.17	0.18
magnetic moment on Mn^{2+} (μ_B)	2.96	2.96
magnetic moment on Mn^{4+} (μ_B)	1.10	1.09

It is worthwhile to note that FM coupled AFM chains were found to be the ground state of another quasi-one-dimensional nitridometalate, $Li_3[FeN_2]$ [44]. Furthermore, the charge-ordered oxomanganates $K_5[Mn_3O_6]$ and $Rb_8[Mn_5O_{10}]$ also show a similar magnetic ordering with respect to the inter- and intra-chain couplings [42]. In these two oxides, the calculated magnetic moments on Mn atoms are close to the spin-only values expected for the high-spin states. Our first principle calculations confirm the charge ordering in $Ba_4[Mn_3N_6]$. QTAIM charges corroborate the oxidation state assignments ($Ba_{ave}^{2+} + 1.29$, $Mn1^{2+} + 0.84$, $Mn2^{4+} + 1.00$, $N_{ave}^{3-} - 1.33$), the low calculated values are in good accordance with other nitridometalates [45]. However, the calculated magnetic moments in the nitridomanganate are reduced by about 2 μ_B for each species in comparison with the anticipated spin-only values of 5.0 μ_B and 3.0 μ_B for high-spin d^5 (Mn^{2+}) and d^3 (Mn^{4+}) configurations, respectively (Table 4). LSDA calculations are known to underestimate magnetic moments in strongly correlated semiconductors. It is also well known that experimentally determined ordered magnetic moments in highly frustrated systems are often reduced in comparison with spin-only values, typically, as a consequence of quantum fluctuations [46]. Therefore, more experimental data would be necessary to probe the importance of on-site correlations in $Ba_4[Mn_3N_6]$.

The electronic density of states for $Ba_4[Mn_3N_6]$ in the AFM1 structure is shown in Figure 6. Well below the Fermi level, the DOS is mainly composed of N(2p) states. A significant contribution of the Mn(3d) states is observed close to the Fermi level, where they get hybridized with the N(2p) states.

Site-resolved DOS contributions from the Mn(3d) states reveal a considerable hybridization between Mn1(3d) and Mn2(3d) states in the region -1.9 eV $< E < 0$ eV, indicating possible Mn–Mn interactions. To study these interactions in more detail, we plotted the COHP curves for all pairs of adjacent Mn atoms in the chains (Figure 7). Below E_F, all Mn–Mn contacts display a predominantly bonding character with the strongest attractive interactions falling in the energy window above -1.9 eV, in consistence with the region of the d-states hybridization. For Mn^{2+}–Mn^{4+} (spin up) and Mn^{4+}–Mn^{4+}, the presence of some anti-bonding states just below E_F reveals a slight under-optimization of bonding, whereas for Mn^{2+}–Mn^{4+} (spin down), the bonding appears to be optimized at E_F. For all pairs, the integrated COHP (ICOHP) values amount to 0.27–0.3 eV bond^{-1} spin direction^{-1}.

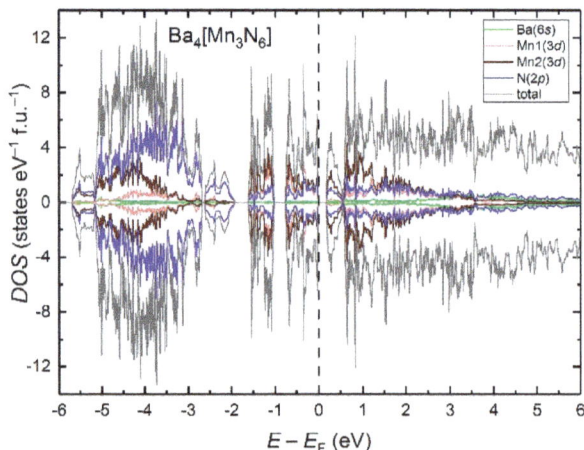

Figure 6. LSDA electronic density of states for Ba$_4$[Mn$_3$N$_6$] in the AFM1 structure. Positive and negative DOS values correspond to major and minor spin channels, respectively.

Interestingly, metal-metal bonding was discussed as a reason for quenching of magnetic moments in certain nitridometalates, e.g., Ca$_{12}$[Mn$_{19}$N$_{23}$] [47], Sc[TaN$_2$] [48], Li$_6$Sr$_2$[Mn$_2$N$_6$] [49] and Ca$_6$[Cr$_2$N$_6$]H [10]. If all adjacent Mn atoms in Ba$_4$[Mn$_3$N$_6$] are linked by 2c–2e bonds, the resulting magnetic moment on every Mn atom will be lowered by 2 μ_B in comparison with the spin-only value. This is in line with the magnetic moments obtained from our LSDA calculations. Therefore, it can be speculated that the reduced magnetic moments for the Mn species are intrinsic to Ba$_4$[Mn$_3$N$_6$] and stem from the chemical bonding between Mn atoms (Figure 8). The development of the Mn–Mn bonding in Ba$_4$[Mn$_3$N$_6$] is reflected in a much shorter Mn–Mn distance (2.5 Å) in comparison with that in the structurally similar oxomanganates K$_5$[Mn$_3$O$_6$] and Rb$_8$[Mn$_5$O$_{10}$] with d(Mn-Mn) = 2.7–2.8 Å, which demonstrate the expected magnetic moments on Mn sites [41,42].

Taking into account the distinctive Mn–Mn interactions, the one-dimensional nitridomanganate anion in Ba$_4$[Mn$_3$N$_6$] can be alternatively understood as a chain of chemically bound Mn atoms decorated by nitride ligands. Such a description provides a link between salt-like alkaline-earth-metal-rich nitridometalates [50], like Li$_7$[MnN$_4$] [40] and Ba$_3$[MnN$_3$] [32], and transition-metal-rich nitrides with pronounced metal-metal interactions, like Ca$_{12}$[Mn$_{19}$N$_{23}$] [47] and Mn$_4$N [36] (Figure 9).

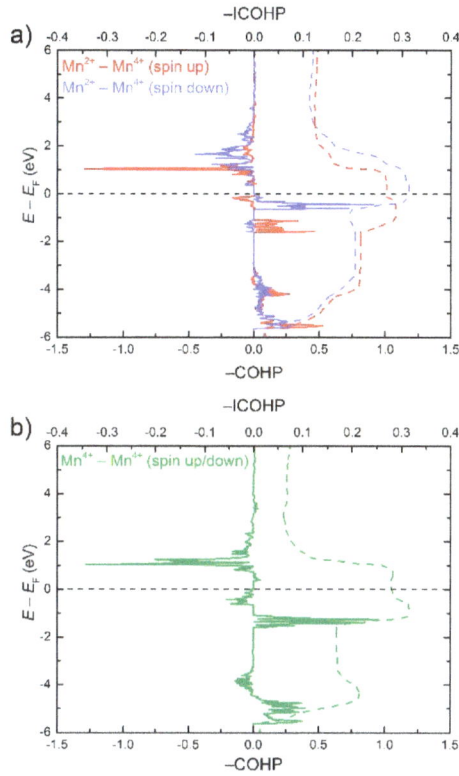

Figure 7. COHP curves for the Mn–Mn nearest neighbor interactions in $Ba_4[Mn_3N_6]$: (**a**) Mn^{2+}–Mn^{4+}; (**b**) Mn^{4+}–Mn^{4+}. For symmetry reasons, the COHP curves for the Mn^{4+}–Mn^{4+} spin up and spin down contacts overlap.

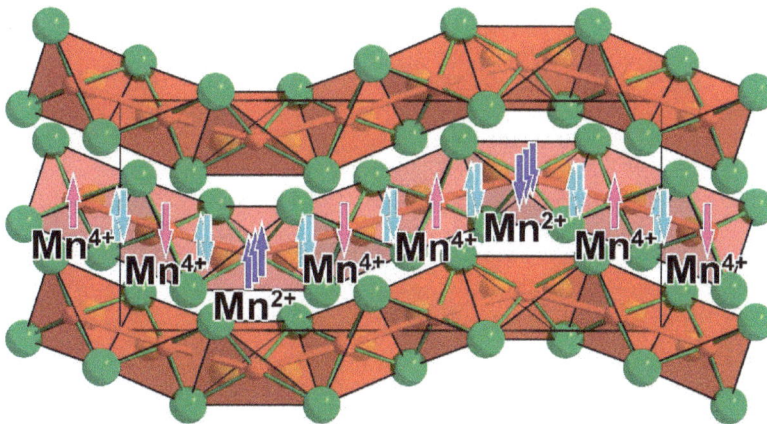

Figure 8. Schematic representation of the possible valence *d*-electron distribution along the chains in $Ba_4[Mn_3N_6]$. Unpaired electrons on Mn^{2+} and Mn^{4+} are shown in blue and pink, respectively. Electron pairs representing Mn–Mn bonds are shown in cyan.

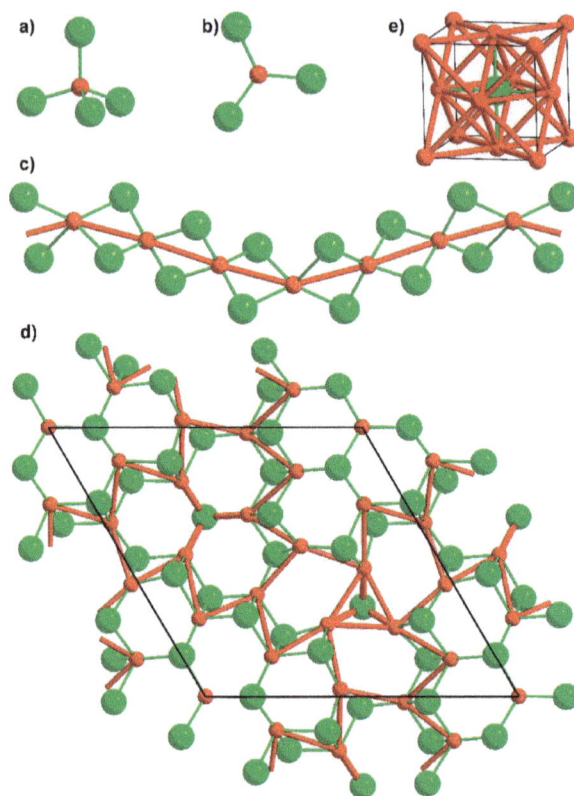

Figure 9. Mn–N and Mn–Mn interactions in nitridomanganates and manganes nitrides from 0D to 3D systems: (**a**) $Li_7[MnN_4]$ [40]; (**b**) $Ba_3[MnN_3]$ [32]; (**c**) $Ba_4[Mn_3N_6]$; (**d**) $Ca_{12}[Mn_{19}N_{23}]$ [47]; (**e**) Mn_4N [36].

4. Conclusions

The mixed-valent nitridomanganate $Ba_4[Mn_3N_6]$ was obtained by high-temperature nitridation of a Ba_2N and Mn mixture. Its crystal structure contains infinite corrugated chains of edge-sharing $[MnN_{4/2}]$ tetrahedra. Analysis of the bond length distribution and first principle calculations suggest a complete charge ordering along the chains with an alternation of Mn^{2+} and Mn^{4+} species in the sequence $[-Mn^{2+}-Mn^{4+}-Mn^{4+}-]$. $Ba_4[Mn_3N_6]$ shows a quasi-one-dimensional magnetic behavior stemming from strong intra-chain and weak inter-chain interactions in this compound, as shown by total energy calculations. In addition, strong bonding interactions were found between Mn atoms along the chains. The emergence of chemical bonding between transition-metal atoms in nitridometalates can provide a link to new structures with extended magnetic topologies.

Supplementary Materials: The following data are available online at http://www.mdpi.com/2073-4352/8/6/235/s1: Results of the chemical analysis, selected bond lengths/angles, DTA-TG curves, high-temperature magnetic susceptibility, and magnetic models used for LSDA calculations. Figure S1. Powder pattern of Ba_2N starting material; Figure S2. DTA-TG curves for a $Ba_4[Mn_3N_6]$ sample (m = 41.79 mg). Heating under flowing Ar (100 mL/min) with a rate of 5 K/min. A slight mass gain in the range T = 300–575 K is probably due to sample oxidation by traces of oxygen; Figure S3. Temperature dependence of magnetic susceptibility for $Ba_4[Mn_3N_6]$ between 320 and 575 K; Figure S4. Model magnetic arrangements used for the total energy calculations (FM—ferromagnetic, AFM—antiferromagnetic, FiM—ferrimagnetic); Table S1. Results of chemical analysis for $Ba_4[Mn_3N_6]$; Table S2. Selected interatomic distances and angles for $Ba_4[Mn_3N_6]$.

Author Contributions: A.O. and P.H. planned and carried out the material synthesis. A.O. and Y.P. planned and carried out the X-ray analysis. A.O., M.B., and W.S. planned and carried out resisitivity and magnetization experiments. A.O. and Y.G. performed quantum chemical calculations. All authors contributed in writing the manuscript. P.H. and Y.G. supervised the project. All authors have given approval to the final version of the manuscript.

Acknowledgments: The authors acknowledge Horst Borrmann and Steffen Hückmann for collecting laboratory PXRD data, Gudrun Auffermann, Anja Völzke, and Ulrike Schmidt for performing chemical analysis, Marcus Schmidt and Susanne Scharsach for conducting DTA/TG measurements, and Ralf Koban for carrying out electrical resistivity measurements. European Synchrotron Radiation Facility (ESRF, Grenoble, France) is acknowledged for providing beamtime at the ID22 beamline managed by Andrew N. Fitch.

Conflicts of Interest: The authors declare no conflict of interest.

References

1. Zhang, W.X.; Ishikawa, R.; Breedlove, B.; Yamashita, M. Single-chain magnets: Beyond the Glauber model. *RSC Adv.* **2013**, *3*, 3772–3798. [CrossRef]

2. Ma, X.S.; Dakic, B.; Naylor, W.; Zeilinger, A.; Walther, P. Quantum simulation of the wavefunction to probe frustrated Heisenberg spin systems. *Nat. Phys.* **2011**, *7*, 399–405. [CrossRef]

3. Bauer, B.; Cincio, L.; Keller, B.P.; Dolfi, M.; Vidal, G.; Trebst, S.; Ludwig, A.W.W. Chiral spin liquid and emergent anyons in a Kagome lattice Mott insulator. *Nat. Commun.* **2014**, *5*, 5137. [CrossRef] [PubMed]

4. Modic, K.A.; Smidt, T.E.; Kimchi, I.; Breznay, N.P.; Biffin, A.; Choi, S.; Johnson, R.D.; Coldea, R.; Watkins-Curry, P.; McCandless, G.T.; et al. Realization of a three-dimensional spin-anisotropic harmonic honeycomb iridate. *Nat. Commun.* **2014**, *5*, 4203. [CrossRef] [PubMed]

5. Pereiro, M.; Yudin, D.; Chico, J.; Etz, C.; Eriksson, O.; Bergman, A. Topological excitations in a Kagome magnet. *Nat. Commun.* **2014**, *5*, 4815. [CrossRef] [PubMed]

6. Van Well, N.; Foyevtsova, K.; Gottlieb-Schönmeyer, S.; Ritter, F.; Manna, R.S.; Wolf, B.; Meven, M.; Pfleiderer, C.; Lang, M.; Assmus, W.; et al. Low-temperature structural investigations of the frustrated quantum antiferromagnets $Cs_2Cu(Cl_{4-x}Br_x)$. *Phys. Rev. B* **2015**, *91*, 035124. [CrossRef]

7. Bisogni, V.; Kourtis, S.; Monney, C.; Zhou, K.J.; Kraus, R.; Sekar, C.; Strocov, V.; Büchner, B.; van den Brink, J.; Braicovich, L.; et al. Femtosecond Dynamics of Momentum-Dependent Magnetic Excitations from Resonant Inelastic X-Ray Scattering in $CaCu_2O_3$. *Phys. Rev. Lett.* **2014**, *112*, 147401. [CrossRef] [PubMed]

8. Pak, C.; Kamali, S.; Pham, J.; Lee, K.; Greenfield, J.T.; Kovnir, K. Chemical Excision of Tetrahedral $FeSe_2$ Chains from the Superconductor FeSe: Synthesis, Crystal Structure, and Magnetism of $Fe_3Se_4(en)_2$. *J. Am. Chem. Soc.* **2013**, *135*, 19111–19114. [CrossRef] [PubMed]

9. Kniep, R.; Höhn, P. 2.06—Low-Valency Nitridometalates. In *Comprehensive Inorganic Chemistry II*, 2nd ed.; Reedijk, J., Poeppelmeier, K., Eds.; Elsevier: Amsterdam, The Netherlands, 2013; pp. 137–160.

10. Bailey, M.S.; Obrovac, M.N.; Baillet, E.; Reynolds, T.K.; Zax, D.B.; DiSalvo, F.J. $Ca_6(Cr_2N_6)H$, the first quaternary nitride–Hydride. *Inorg. Chem.* **2003**, *42*, 5572–5578. [CrossRef] [PubMed]

11. Höhn, P.; Ballé, J.T.; Fix, M.; Prots, Y.; Jesche, A. Single Crystal Growth and Anisotropic Magnetic Properties of $Li_2Sr[Li_{1-x}Fe_xN]_2$. *Inorganics* **2016**, *4*, 4040042. [CrossRef]

12. STOE & Cie GmbH. *WinXPow*; STOE & Cie GmbH: Darmstadt, Germany, 2003.

13. Altomare, A.; Camalli, M.; Cuocci, C.; Giacovazzo, C.; Moliterni, A.; Rizzi, R. EXPO2009: Structure solution by powder data in direct and reciprocal space. *J. Appl. Crystallogr.* **2009**, *42*, 1197–1202. [CrossRef]

14. Petřiček, V.; Dušek, M.; Palatinus, L. Crystallographic Computing System JANA2006: General features. *Z. Kristallogr.* **2014**, *229*, 345–352. [CrossRef]

15. Honda, K. Die thermomagnetischen Eigenschaften der Elemente. *Ann. Phys.* **1910**, *337*, 1027–1063. [CrossRef]

16. Koepernik, K.; Eschrig, H. Full-potential nonorthogonal local-orbital minimum-basis band-structure scheme. *Phys. Rev. B* **1999**, *59*, 1743–1757. [CrossRef]

17. Jepsen, O.; Burkhardt, A.; Andersen, O.K. *The Stuttgart Tight-Binding LMTO-ASA Program*; Version 4.7; Max-Planck-Institut für Festkörperforschung: Stuttgart, Germany, 1998.

18. Perdew, J.P.; Wang, Y. Accurate and simple analytic representation of the electron-gas correlation energy. *Phys. Rev. B* **1992**, *45*, 13244–13249. [CrossRef]

19. Von Barth, U.; Hedin, L. Local exchange-correlation potential for spin polarized case: 1. *J. Phys. C Solid State* **1972**, *5*, 1629–1642. [CrossRef]

20. Andersen, O.K. Linear Methods in Band Theory. *Phys. Rev. B* **1975**, *12*, 3060–3083. [CrossRef]
21. Dronskowski, R.; Blöchl, P.E. Crystal orbital Hamilton populations (COHP): Energy-resolved visualization of chemical bonding in solids based on density-functional calculations. *J. Phys. Chem.* **1993**, *97*, 8617–8624. [CrossRef]
22. Kohout, M. *DGRID 4.6*; Springer: Radebeul, Germany, 2011.
23. Bader, R.F.W. *Atoms in Molecules—A Quantum Theory*, 2003 ed.; Clarendon Press: Oxford, UK, 1994; Volume 22, pp. 1–456.
24. Ludwig, M.; Niewa, R.; Kniep, R. Dimers $[Al_2N_6]^{12-}$ and chains $^1_\infty[AlN_{4/2}^{3-}]$ in the crystal structures of $Ca_6[Al_2N_6]$ and $Ba_3[Al_2N_4]$. *Z. Naturforsch. B* **1999**, *54*, 461–465. [CrossRef]
25. Blase, W.; Cordier, G.; Ludwig, M.; Kniep, R. $Sr_3[Al_2N_4]$: Ein Nitridoaluminat mit gewellten Tetraederketten $^1_\infty[AlN_{4/2}^{3-}]$. *Z. Naturforsch. B* **1994**, *49*, 501–505. [CrossRef]
26. Clarke, S.J.; DiSalvo, F.J. Synthesis and structure of one-, two-, and three-dimensional alkaline earth metal gallium nitrides: $Sr_3Ga_2N_4$, $Ca_3Ga_2Ne_4$, and $Sr_3Ga_3N_5$. *Inorg. Chem.* **1997**, *36*, 1143–1148. [CrossRef] [PubMed]
27. Yamane, H.; DiSalvo, F.J. $Ba_3Ga_2N_4$. *Acta Crystallogr. C* **1996**, *52*, 760–761. [CrossRef]
28. Park, D.G.; Gál, Z.A.; DiSalvo, F.J. Sr_3GeMgN_4: New quaternary nitride containing Mg. *J. Alloy. Compd.* **2003**, *360*, 85–89. [CrossRef]
29. Park, D.G.; DiSalvo, F.J. A Structural Comparison between a New Quaternary Nitride, Ba_3GeMgN_4, and Its Isostructural Sr analogue. *Bull. Korean Chem. Soc.* **2011**, *32*, 353–355. [CrossRef]
30. Gudat, A.; Kniep, R.; Rabenau, A.; Bronger, W.; Ruschewitz, U. Li_3FeN_2, a ternary nitride with $^1_\infty[FeN^{3-}_{4/2}]$ chains: Crystal structure and magnetic properties. *J. Less Common Met.* **1990**, *161*, 31–36. [CrossRef]
31. Bérar, J.F.; Lelann, P. E.s.d.s and estimated probable-error obtained in Rietveld refinements with local correlations. *J. Appl. Crystallogr.* **1991**, *24*, 1–5. [CrossRef]
32. Tennstedt, A.; Röhr, C.; Kniep, R. $Sr_3[MnN_3]$ and $Ba_3[MnN_3]$, the First Nitridomanganates(III): Trigonal-Planar Anions $[Mn^{III}N_3]^{6-}$. *Z. Naturforsch. B* **1993**, *48*, 794–796. [CrossRef]
33. Yamada, T.; Fujii, Y. The Crystal Structure of α-Mn Reexamined on Single Crystal Specimens. *J. Phys. Soc. Jpn.* **1970**, *28*, 1503–1507. [CrossRef]
34. Shoemaker, C.B.; Shoemaker, D.P.; Hopkins, T.E.; Yindepit, S. Refinement of the structure of b-manganese and of a related phase in the Mn-Ni-Si system. *Acta Crystallogr. B* **1978**, *34*, 3573–3576. [CrossRef]
35. Aoki, M.; Yamane, H.; Shimada, M.; Kajiwara, T. Single crystal growth of Mn_2N using an In-Na flux. *Mater. Res. Bull.* **2004**, *39*, 827–832. [CrossRef]
36. Juza, R.; Deneke, K.; Puff, H. Ferrimagnetismus der Mischkristalle von Mn_4N mit Chrom, Eisen und Nickel—41. Mitteilung über Metallnitride und -amide. *Z. Elektrochem.* **1959**, *63*, 551–557.
37. Brese, N.E.; O'Keeffe, M. Bond-valence parameters for solids. *Acta Crystallogr. B* **1991**, *47*, 192–197. [CrossRef]
38. Wintenberger, M.; Guyader, J.; Maunaye, M. Étude cristallographique et magnétique de $MnGeN_2$ par diffraction neutronique. *Solid State Commun.* **1972**, *11*, 1485–1488. [CrossRef]
39. Grins, J.; Käll, P.O.; Svensson, G. Synthesis and structural characterisation of $MnWN_2$ prepared by ammonolysis of $MnWO_4$. *J. Mater. Chem.* **1995**, *5*, 571–575. [CrossRef]
40. Niewa, R.; Wagner, F.R.; Schnelle, W.; Hochrein, O.; Kniep, R. $Li_{24}[MnN_3]_3N_2$ and $Li_5[(Li_{1-x}Mn_x)N]_3$, Two Intermediates in the Decomposition Path of $Li_7[MnN_4]$ to $Li_2[(Li_{1-x}Mn_x)N]$: An Experimental and Theoretical Study. *Inorg. Chem.* **2001**, *40*, 5215–5222. [CrossRef] [PubMed]
41. Coey, J.M.D.; Viret, M.; von Molnar, S. Mixed-valence manganites. *Adv. Phys.* **1999**, *48*, 167–293. [CrossRef]
42. Nuss, J.; Dasari, P.L.V.K.; Jansen, M. $K_5Mn_3O_6$ and $Rb_8Mn_5O_{10}$, New Charge Ordered Quasi One-Dimensional Oxomanganates (II, III). *Z. Anorg. Allg. Chem.* **2015**, *641*, 316–321. [CrossRef]
43. Michel, C.; Baranovskii, S.D.; Klar, P.J.; Thomas, P.; Goldlucke, B. Strong non-Arrhenius temperature dependence of the resistivity in the regime of traditional band transport. *Appl. Phys. Lett.* **2006**, *89*, 112116. [CrossRef]
44. Ming, X.; Wang, X.L.; Du, F.; Han, B.; Wang, C.Z.; Chen, G. Unusual intermediate spin Fe^{3+} ion in antiferromagnetic Li_3FeN_2. *J. Appl. Phys.* **2012**, *111*, 063704. [CrossRef]
45. Bronger, W.; Baranov, A.; Wagner, F.R.; Kniep, R. Atom Volumina and Charge Distributions in Nitridometalates. *Z. Anorg. Allg. Chem.* **2007**, *633*, 2553–2557. [CrossRef]
46. Jiang, H.C.; Kruger, F.; Moore, J.E.; Sheng, D.N.; Zaanen, J.; Weng, Z.Y. Phase diagram of the frustrated spatially-anisotropic $S = 1$ antiferromagnet on a square lattice. *Phys. Rev. B* **2009**, *79*, 174409. [CrossRef]

47. Ovchinnikov, A.; Bobnar, M.; Prots, Y.; Borrmann, H.; Sichelschmidt, J.; Grin, Y.; Höhn, P. $Ca_{12}[Mn_{19}N_{23}]$ and $Ca_{133}[Mn_{216}N_{260}]$: Structural complexity by 2D intergrowth. *Angew. Chem.* **2018**. [CrossRef]

48. Niewa, R.; Zherebtsov, D.A.; Schnelle, W.; Wagner, F.R. Metal-metal bonding in $ScTaN_2$. A new compound in the system ScN-TaN. *Inorg. Chem.* **2004**, *43*, 6188–6194. [CrossRef] [PubMed]

49. Bolvin, H.; Wagner, F.R. Case of a Strong Antiferromagnetic Exchange Coupling Induced by Spin Polarization of a Mn-Mn Partial Single Bond. *Inorg. Chem.* **2012**, *51*, 7112–7118. [CrossRef] [PubMed]

50. Höhn, P.; Niewa, R. Nitrides of Non-Main Group Elements. In *Handbook of Solid State Chemistry, Part 1. Materials and Structure of Solids*; Wiley-VCH Verlag GmbH & Co. KGaA: Weinheim, Germany, 2017; pp. 251–359.

crystals

MDPI

Article

Optimization of $Ca_{14}MgSb_{11}$ through Chemical Substitutions on Sb Sites: Optimizing Seebeck Coefficient and Resistivity Simultaneously

Yufei Hu, Kathleen Lee and Susan M. Kauzlarich *

Department of Chemistry, University of California, One Shields Avenue, Davis, CA 95616, USA;
yfhu@ucdavis.edu (Y.H.); Kathy.Lee@jpl.nasa.gov (K.L.)
* Correspondence: smkauzlarich@ucdavis.edu

Received: 17 April 2018; Accepted: 8 May 2018; Published: 13 May 2018

Abstract: In thermoelectric materials, chemical substitutions are widely used to optimize thermoelectric properties. The *Zintl* phase compound, $Yb_{14}MgSb_{11}$, has been demonstrated as a promising thermoelectric material at high temperatures. It is *iso*-structural with $Ca_{14}AlSb_{11}$ with space group $I4_1/acd$. Its *iso*-structural analog, $Ca_{14}MgSb_{11}$, was discovered to be a semiconductor and have vacancies on the Sb(3) sites, although in its nominal composition it can be described as consisting of fourteen Ca^{2+} cations with one $[MgSb_4]^{9-}$ tetrahedron, one Sb_3^{7-} linear anion and four isolated Sb^{3-} anions (Sb(3) site) in one formula unit. When Sn substitutes Sb in $Ca_{14}MgSb_{11}$, optimized Seebeck coefficient and resistivity were achieved simultaneously although the Sn amount is small (<2%). This is difficult to achieve in thermoelectric materials as the Seebeck coefficient and resistivity are inversely related with respect to carrier concentration. Thermal conductivity of $Ca_{14}MgSb_{11-x}Sn_x$ remains almost the same as $Ca_{14}MgSb_{11}$. The calculated zT value of $Ca_{14}MgSb_{10.80}Sn_{0.20}$ reaches 0.49 at 1075 K, which is 53% higher than that of $Ca_{14}MgSb_{11}$ at the same temperature. The band structure of $Ca_{14}MgSb_7Sn_4$ is calculated to simulate the effect of Sn substitutions. Compared to the band structure of $Ca_{14}MgSb_{11}$, the band gap of $Ca_{14}MgSb_7Sn_4$ is smaller (0.2 eV) and the Fermi-level shifts into the valence band. The absolute values for density of states (DOS) of $Ca_{14}MgSb_7Sn_4$ are smaller near the Fermi-level at the top of valence band and $5p$-orbitals of Sn contribute most to the valence bands near the Fermi-level.

Keywords: Zintl; $Ca_{14}AlSb_{11}$; polar intermetallic; thermoelectric

1. Introduction

Thermoelectric materials have attracted significant attention as they can improve the efficiency of energy through converting wasted heat into electricity. The efficiency of thermoelectric materials can be evaluated through the figure of merit (zT) by Equation (1).

$$zT = \frac{\alpha^2}{\rho\kappa}T \tag{1}$$

In the equation, α is the Seebeck coefficient, ρ is electrical resistivity, T is the absolute temperature and κ is thermal conductivity. Defects or tiny amounts of chemical substitutions are important to the optimizations of thermoelectric properties in some typical thermoelectric materials as defects can tune carrier concentrations effectively and adjust the Seebeck coefficient and electrical resistivity. Cationic defects, tuned carrier concentrations and optimized thermoelectric properties were observed in *Zintl* phase compounds [1–3]. Defects can also scatter phonons, make the systems phonon-glass-like and decrease lattice thermal conductivity. The Zn defects in Zn-Sb compounds and Cu defects in Cu-chalcogenides are essential for decreasing lattice thermal conductivity and tuning of carrier concentrations [4–10]. In some Type-I clathrates, defects have been found in frameworks and the ordering of vacancies are important for the tuning of the Seebeck coefficient, electrical resistivity and thermal conductivity [11–15]. Defects in clathrates can also change the band structure as the bonding and antibonding orbitals of framework elements contribute to the bands near the Fermi-level [11].

Considering the fact that defects are actually replacing atoms with voids, small amounts of chemical substitutions also have similar effects on thermoelectric properties. 1.25% Zn substitutions of IIIA atoms in *p*-type thermoelectric materials $Eu/Sr_5In_2Sb_6$ and $BaGa_2Sb_2$ add a hole in the structure and are able to significantly change the carrier concentrations, therefore changing electrical resistivity and Seebeck coefficient [16–18]. In IV-VI materials, 1–2% of potassium, sodium or thulium doping can improve the thermoelectric properties significantly [19–23]. The benefit of K or Na doping may reside in two aspects. The first one is the change of DOS near the Fermi-level, which can increase the effective mass and lead to an improved Seebeck coefficient. The other benefit is the formation of a *nano*-composite, which can effectively decrease lattice thermal conductivity. More recently, a GeTe-based material with Pb and Bi_2Te_3 doping (3 mol %) was discovered to have a *zT* value of 1.9 at 773K [24]. 1% doping of rare earth elements in TAGS-85 samples increases Seebeck coefficient and therefore leads to larger *zT* values [25,26]. In half-*Heusler* alloys, small amounts of substitutions (1%) can also dramatically (~50%) improve thermoelectric properties [27–30].

The *Zintl* phase compounds, $Yb_{14}MgSb_{11}$ and $Ca_{14}MgSb_{11}$, were reported in 2014 and their thermoelectric properties studied [31]. Both of them have the $Ca_{14}AlSb_{11}$-type of structure (Figure 1 and Table 1), and $Ca_{14}MgSb_{11}$ was found to have vacancies (2.6%) on the isolated Sb(3) sites. These sites are coordinated by Ca^{2+} cations and are shown as the non-bonded Sb atoms in Figure 1. $Yb_{14}MgSb_{11}$ has a *zT* of ~1 at 1075 K while $Ca_{14}MgSb_{11}$ has a semiconductor-like resistivity and a *zT* of 0.32 at 1075 K. Their *iso*-structural analog, $Yb_{14}MnSb_{11}$, has also been discovered as a good thermoelectric material and many studies have been conducted to optimize its thermoelectric properties [2,32]. Samples synthesized by powder metallurgy have slightly larger (~20 Å^3 out of ~6000 Å^3) unit cell parameters than crystals synthesized by the Sn-flux method based on the refinement of powder X-ray diffraction patterns. This suggests that crystals of $Yb_{14}MnSb_{11}$ synthesized by the Sn-flux method may have small amounts of Yb or Mn vacancies. The smaller resistivity and lower Seebeck coefficient of $Yb_{14}MnSb_{11}$ samples synthesized as crystals are also consistent with the above statements. Further studies show that 1% Te or Ge substitutions on Sb site can significantly alter the thermoelectric properties [33,34].

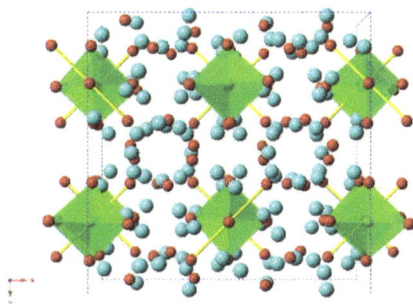

Figure 1. Unit cell of $Ca_{14}MgSb_{11}$ projected along the *c*-axis. Ca and Sb atoms are represented by blue and red spheres, Sb_3^{7-} ions are shown with yellow bonds and the green tetrahedra are $[MgSb_4]^{9-}$ clusters.

The synthesis and thermoelectric properties of $Ca_{14}MgSb_{11-x}Sn_x$ are systematically investigated. Sn is used to compensate for Sb vacancies and thereby improve the overall thermoelectric properties of $Ca_{14}MgSb_{11}$. Sn has one electron less than Sb, which will tune the carrier concentration and resistivity.

Calculations of the electronic band structures show that Sb(3) sites contribute most near the Fermi-level and therefore substitution of this site may dramatically change the thermoelectric properties [31].

Table 1. Wyckoff positions and atomic coordinates [$\times 10^4$] of atoms for $Ca_{14}MgSb_{11}$.

Atom	Wyckoff Positions	x	y	z
Ca1	32g	9580(2)	9274(2)	8281(1)
Ca2	32g	9771(2)	1264(2)	79(1)
Ca3	16e	3553(2)	0	2500
Ca4	32g	1781(2)	4030(2)	8439(1)
Mg1	8a	0	2500	8750
Sb1	16f	1364(1)	3864(1)	1250
Sb2	32g	37(1)	1100(1)	8059(1)
Sb3 *	32g	8705(1)	9751(1)	9516(1)
Sb4	8b	0	2500	1250

* Sb3 site has 2.6% vacancy.

2. Experimental Section

Reagents. Elemental Ca pieces (99.5%, Alfa Aesar, Tewksbury, MA, USA), Mg turnings (99.98%, Sigma-Aldrich, St. Louis, MO, USA), Sb (99.999%, Alfa Aesar, Tewksbury, MA, USA) and Sn (99.3%, Alfa Aesar, Tewksbury, MA, USA) were used for the synthesis. Ca was cut into small pieces while Mg and Sn were used as received. All elements were handled using inert atmosphere techniques, including an argon filled glovebox with water levels <0.5 ppm.

Synthesis of Powder. Quantitative yield, high purity samples were synthesized through a powder metallurgy method [2]. Melting Sn and Mg together at 600 °C with the ratio 1:2.2 to produce Mg_2Sn as precursor. Samples with the ratio Ca:Mg:Sb:Mg_2Sn = 14:1.1 $\times$ (1−2x):11−x:x (x is the Sn amount) were loaded into a 50 cm^3 tungsten carbide ball mill vial with one large WC ball (diameter = 11 mm) and two small WC balls (diameter = 8 mm). The mixtures of elements were ball milled on a SPEX 8000 M (SPEX SamplePrep, Metuchen, NJ, USA) for one hour and another 30 min after a 30 min break (the 30 min break prevents the reaction mixture from becoming too hot). The fine powder was transferred to a glovebox and then into a niobium tube, which was sealed by arc welding under argon and further jacketed under vacuum in fused silica. $Ca_{14}MgSb_{11-x}Sn_x$ is annealed at 800 °C for 4 days with a heating rate of 30 °C/h [31].

Powder X-ray Diffraction. Samples were examined using a Bruker zero background holder on a Bruker D8 Advance Diffractometer operated at 40 kV and 40 mA utilizing Cu $K\alpha$ radiation. $K\beta$ radiation is removed by a Ni filter. WinPLOTR (version Jan 2012, part of the FullProf suite of programs, https://www.ill.eu/sites/fullprof/ University of Rennes 1, Rennes, France) software was used for background subtraction and pattern analysis, and EDPCR 2.00 software (part of the FullProf suite of programs, https://www.ill.eu/sites/fullprof/ University of Rennes 1, Rennes, France) was used to perform Le Bail refinement [35].

Consolidation of Powder. The bulk powder samples were consolidated into dense pellets via a Dr. Sinter Lab Jr. SPS-211Lx or SPS-2050 spark plasma sintering (SPS) system (Sumitomo, Tokyo, Japan) in a 12.7 mm high-density graphite die (POCO) under vacuum (<10 Pa). The temperature was increased from room temperature to 1000–1025 K in 15 min, and remained stable for 5–15 min. When the temperature reached the maximum, the force loaded increased from 3 kN to 5–8 kN. The samples were cooled to room temperature afterwards. The geometrical sample density was larger than 95% of the theoretical density.

Thermal Conductivity. Thermal diffusivity (D) measurement was conducted on the pellet obtained from SPS from 300 K to 1075 K on a Netzsch LFA-457 laser flash unit (Netzsch, Burlington, MA, USA). The pellet surfaces were well polished and coated with graphite. The measurement was conducted under dynamic argon atmosphere with a flow rate of 50 mL/min. Thermal conductivity was calculated using the equation $\kappa = D \times \rho \times C_p$. Room-temperature density was measured from a volume method and the high-temperature density was derived using thermal expansion data from a previous paper on $Yb_{14}MnSb_{11}$ [36]. The C_p was taken from previous papers, which was measured by differential scanning calorimetry (DSC) [2].

Electrical Transport Properties. A Linseis LSR-3 unit (Linseis, Robinsville, NJ, USA) was employed to measure Seebeck coefficient and electrical resistivity via a four-probe method from 325 K to 1075 K under a helium atmosphere on a bar-shaped sample [37]. The sample which had been previously measured on the LFA instrument was cut into a $2 \times 2 \times 11$ mm bar using a Buehler diamond saw and polished before measurement. The probe distance was 8 mm. For convenience and clarity, Seebeck coefficient, electrical resistivity and thermal conductivity were fit to six-order polynomial functions to calculate zT values. Room-temperature Hall coefficient was measured with a Quantum Design physical property measurement system (PPMS) from 7 T to -7 T by 5-point *ac* technique. Platinum leads were connected to the pressed pellet through silver paste. Carrier concentration was calculated using equation $R_H = -1/ne$ using the average of R_H from different magnetic fields.

Quantum-chemical calculations. Density functional band structure calculations for $Ca_{14}MgSb_{11}$ and $Ca_{14}MgSb_7Sn_4$ were performed using the linear-muffin tin orbital method (TB-LMTO, Stuttgart, Germany, version 47.1b) within the tight binding approximation [38–42]. The density of states (DOS) and band structures were calculated after convergence of the total energy on a dense k-mesh with $12 \times 12 \times 12$ points, with 163 irreducible k-points. A basis set containing Ca(4*s*,3*d*), Mg(3*s*,3*p*), and Sb(5*s*,5*p*) orbitals was employed for a self-consistent calculation, with Ca(4*p*), Mg(3*d*) and Sb(5*d*,4*f*) functions being downfolded.

3. Results and Discussion

Synthesis and Structure. Mg_2Sn used in the reaction as a precursor was verified by powder X-ray diffraction. There may be very small peaks indicating the existence of unreacted Mg since extra Mg is used in the precursor synthesis. As mentioned in a previous paper, $Ca_{14}MgSb_{11}$ synthesized by powder metallurgy may contain a minor amount of impurity $Ca_{11}Sb_{10}$ and its existence has a limited effect on the thermoelectric properties [31,43]. $Ca_{14}MgSb_{11-x}Sn_x$ ($x = 0.05$ and 0.10) also contained minor amounts of impurities $Ca_{11}Sb_{10}$ while the Le Bail refinement of the powder X-ray diffraction pattern for $Ca_{14}MgSb_{10.80}Sn_{0.20}$ indicates a phase pure sample (Figure 2). These samples are air-sensitive and oxidized rapidly upon exposure, leading to poor quality of powder X-ray diffraction patterns, not suitable for Rietveld refinement. The refined unit cell parameters are listed in Table 2. Generally, the unit cell parameters show a slight increase with the increasing Sn amounts.

Table 2. Unit cell parameters from refinement of powder X-ray diffraction patterns.

Sn Amount Used in Synthesis	*a* (Å)	*c* (Å)	*V* (Å³)
0.00	16.73(1)	22.54(1)	6309(1)
0.05	16.72(1)	22.60(1)	6318(3)
0.10	16.73(1)	22.59(1)	6323(3)
0.20	16.73(1)	22.62(1)	6331(3)

Thermoelectric Properties. Figure 3 shows the results of thermoelectric properties measurement of $Ca_{14}MgSb_{10.95}Sn_x$ ($x = 0.05$, 0.1 and 0.2). The Seebeck coefficient and resistivity of these samples change dramatically with the small changes of Sn compositions. The Seebeck coefficient of $Ca_{14}MgSb_{10.95}Sn_{0.05}$ has a similar trend to that of $Ca_{14}MgSb_{11}$, but the values are much lower. $Ca_{14}MgSb_{10.90}Sn_{0.10}$ and $Ca_{14}MgSb_{10.80}Sn_{0.20}$ have almost the same Seebeck coefficient values, which are slightly higher than the Seebeck coefficient of $Ca_{14}MgSb_{11}$. The linear increase of Seebeck coefficients of $Ca_{14}MgSb_{10.90}Sn_{0.10}$ and $Ca_{14}MgSb_{10.80}Sn_{0.20}$ within the measured temperature region is the most significant change caused by Sn substitutions. The Seebeck coefficient of $Ca_{14}MgSb_{11}$, which decreases at low temperature and increases at high temperature, is attributed to a combination of increasing temperature and change of carrier concentration based on Equation (2) [19,31].

$$\alpha = \frac{8\pi^2 k_B^2}{3eh^2} m^* T \left(\frac{\pi}{3n}\right)^{\frac{2}{3}} \qquad (2)$$

In the equation, k_B is the Boltzmann constant, h is the Planck constant, m^* is the effective mass of carriers and n is carrier concentration. In $Ca_{14}MgSb_{10.90}Sn_{0.10}$ and $Ca_{14}MgSb_{10.80}Sn_{0.20}$, no effect of carrier concentration change is observed in the Seebeck coefficient, which indicates that the substitution of Sn decreases the band gap.

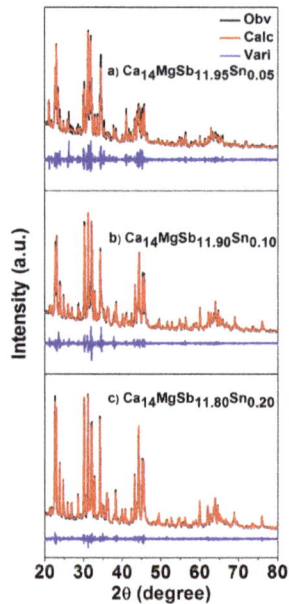

Figure 2. Refinement of the powder X-ray diffraction patterns of (a) $Ca_{14}MgSb_{10.95}Sn_{0.05}$, (b) $Ca_{14}MgSb_{10.90}Sn_{0.10}$ and (c) $Ca_{14}MgSb_{10.80}Sn_{0.20}$.

The resistivity of these samples are semiconductor-like, but the details are different. $Ca_{14}MgSb_{10.95}Sn_{0.05}$ has a larger resistivity than that of $Ca_{14}MgSb_{11}$ at room temperature and a 33% lower resistivity than $Ca_{14}MgSb_{11}$ at high temperatures. The resistivity of $Ca_{14}MgSb_{10.90}Sn_{0.10}$ is higher than the other two samples, especially in the high temperature region. The resistivity of $Ca_{14}MgSb_{10.80}Sn_{0.20}$ follows the expectation, which is lower than that of $Ca_{14}MgSb_{11}$ for the entire temperature region. The activation energy (E_a) can be calculated based on Equation (3) and are listed in Table 3 [31].

$$ln\,\rho = ln\,\rho_0 + E_a/2k_B T \qquad (3)$$

Table 3. Calculated activation energy of $Ca_{14}MgSb_{11-x}Sn_x$.

Composition	Activation Energy (eV)
$x = 0.00$	0.15
$x = 0.05$	0.17
$x = 0.10$	0.08
$x = 0.20$	0.10

It can be seen that the activation energy drops from $x = 0.05$ to $x = 0.10$. The carrier concentration of $Ca_{14}MgSb_{10.80}Sn_{0.20}$ is measured to be $3.5 \times 10^{19}/cm^3$ at room temperature, which is slightly higher than that of $Ca_{14}MgSb_{11}$. Therefore, the carrier concentration has a small change within the three compositions. The Sn substitution of Sb is expected to make $Ca_{14}MgSb_{11}$ more conductive as Sn has one electron less than Sb and this increases the carrier concentration of p-type semiconductor $Ca_{14}MgSb_{11}$. However, the carrier concentration change is complex due to the combinations of defects, hole additions caused by Sn doping and the band structure change. The addition of Sn changes the activation energy, which corresponds to the band gap, but this cannot change the intrinsic semiconductor properties of $Ca_{14}MgSb_{11}$.

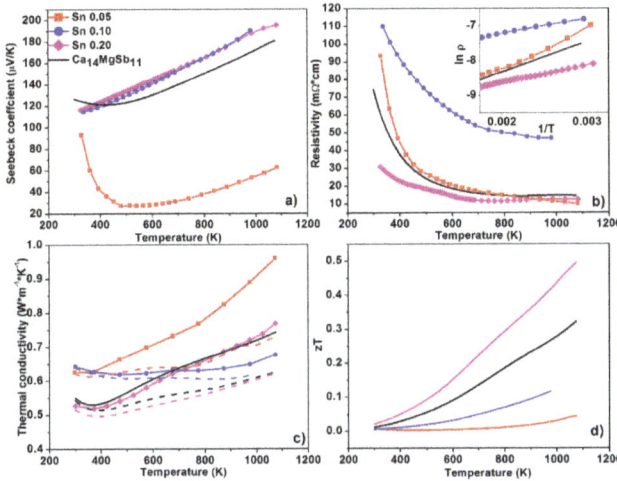

Figure 3. (**a**) Seebeck coefficient, (**b**) resistivity (insert: *ln* ρ vs. $1/T$ plots), (**c**) thermal conductivity with lattice thermal conductivities shown in dashed lines and (**d**) zT values of $Ca_{14}MgSb_{10.95}Sn_{0.05}$, $Ca_{14}MgSb_{10.90}Sn_{0.10}$ and $Ca_{14}MgSb_{10.80}Sn_{0.20}$. In all plots, data for $Ca_{14}MgSb_{11}$ are shown in black curves as references.

The thermal conductivities of $Ca_{14}MgSb_{11-x}Sn_x$ ($x = 0.05, 0.10$ and 0.20) are listed in Figure 2c. The sample of $x = 0.05$ has the largest thermal conductivity and sample of $x = 0.20$ has almost the same thermal conductivity as $Ca_{14}MgSb_{11}$. Lorenz numbers of these samples at different temperatures are calculated using Equations (4) and (5) [4,44].

$$L = \left(\frac{k_B}{e}\right)^2 \frac{3F_0(\eta)F_2(\eta) - 4F_1(\eta)^2}{F_0(\eta)^2} \tag{4}$$

$$\alpha = \frac{k_B}{e}\left[\frac{2F_1(\eta)}{F_0(\eta)} - \eta\right] \tag{5}$$

$$\kappa_{lat} = \kappa - \kappa_e = \kappa - L\sigma T \tag{6}$$

In the equations, L is the Lorenz number, η is the reduced Fermi-level, $F_n(\eta)$ is the Fermi-Dirac integral, κ_{lat} is the lattice thermal conductivity and κ_e is electronic thermal conductivity from electrical conductivity. The lattice thermal conductivity can be calculated by subtracting the electronic thermal conductivity from the total thermal conductivity. The calculated lattice thermal conductivity shows that the lattice term contributes the most to the total thermal conductivity and the electronic term has larger contributions at high temperature regions due to the decrease of resistivity. The calculated zT values show that $Ca_{14}MgSb_{10.80}Sn_{0.20}$ has the largest maximum zT value of 0.49 at 1075 K, which is

53% higher than that of $Ca_{14}MgSb_{11}$ due to optimized resistivity. $Ca_{14}MgSb_{10.95}Sn_{0.05}$ has a maximum zT value of 0.04 at 1075 K due to a lower Seebeck coefficient and $Ca_{14}MgSb_{10.90}Sn_{0.10}$ has a maximum zT value of 0.12 due to a much larger resistivity.

4. DOS Calculation

The DOS diagram for $Ca_{14}MgSb_7Sn_4$ is shown in Figure 4. This composition is assumed for calculating the DOS of Sn-substituted $Ca_{14}MgSb_{11}$, as the Sb(3) site is assumed to be occupied by Sn. Several reasons support this hypothesis as the Sb(3) site is known to show vacancies and Sb(3) is the isolated Sb^{3-} site with no direct covalent bonding to other atoms, which is the easiest to be substituted. Because of the large unit cell of $Ca_{14}MgSb_{11}$, Sb(3) is assumed to be fully replaced by Sn although actually at most only 5% of Sb(3) is occupied by Sn. Therefore, the composition used in the calculations is $Ca_{14}MgSb_7Sn_4$.

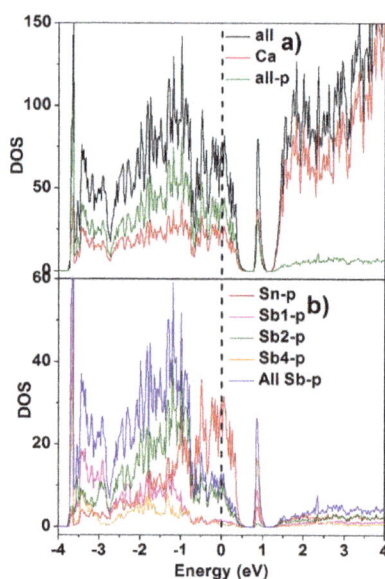

Figure 4. Density of states (DOS) diagram for $Ca_{14}MgSb_7Sn_4$. (**a**) Total DOS together with partial contributions from Ca and 5*p*-orbitals of Sn and Sb (**b**) Partial DOS showing the contributions of only 5*p*-orbitals of Sn and Sb.

The band structure of $Ca_{14}MgSb_7Sn_4$ (Figure 4) has similar features to that of $Ca_{14}MgSb_{11}$ [31]. The Fermi-level falls into the valence band, to which *p*-orbitals of Sb and Sn have dominating contributions. Sn makes the largest contribution at the top of the valence band near the Fermi-level. The conduction band is mainly from orbitals of Ca, especially when energy is >1.2 eV, and there is a sharp peak at the bottom of the conduction band, which originates from the anti-bonding orbitals of the linear Sb_3^{7-} units and orbitals of Ca. The band gap between the top of the valence band (+0.58 eV) and the bottom of the conduction band (+0.78 eV) is 0.20 eV. This value is smaller than the band gap of $Ca_{14}MgSb_{11}$ (0.6 eV) but is similar to the activation energy calculated from resistivity. It is also noticeable that there is an energy gap of 0.15 eV between the sharp peak and the states at higher energy in the conduction band. Compared to $Ca_{14}MgSb_{11}$, $Ca_{14}MgSb_7Sn_4$ has smaller absolute values for the density of states at the top of valence band near the Fermi-level. Therefore, when Sb is replaced by Sn, two major changes happen to the valence band. The first one is the shift of Fermi-level into the valence band and the other one is the decrease of absolute values of DOS near Fermi-level.

In the experimental samples $Ca_{14}MgSb_{11-x}Sn_x$ (x = 0.05, 0.10 and 0.20), the Sn amount is much lower than that used in the calculations for the DOS. Comparing the electronic calculations for $Ca_{14}MgSb_{11}$ and $Ca_{14}MgSb_7Sn_4$, a small decrease of band gap is expected. This is supported by the experimental results. The Seebeck coefficient strongly depends on the density of states at the Fermi-level [19,45,46].

$$S = \frac{\pi^2}{3} \frac{k_B}{q} k_B T \left[\frac{1}{n} \frac{dn(E)}{dE} + \frac{1}{\mu} \frac{d\mu(E)}{dE} \right]_{E=E_F} \tag{7}$$

$$S = \frac{\pi^2}{3} \frac{k_B}{q} k_B T \left\{ \frac{1}{n} \frac{d[g(E) * f(E)]}{dE} + \frac{1}{\mu} \frac{d\mu(E)}{dE} \right\}_{E=E_F} \tag{8}$$

In Equations (7) and (8), μ is the mobility, n is the carrier density, $g(E)$ is density of states and $f(E)$ is Fermi function [19]. In $Ca_{14}MgSb_{11-x}Sn_x$ (x = 0.05, 0.10 and 0.20), as mentioned above, the Fermi level shifts into valence band and the change of DOS at the Fermi level is difficult to determine. Based on the measured Seebeck coefficient, it can be concluded that Fermi-level of $Ca_{14}MgSb_{10.95}Sn_{0.05}$ falls into a valley of DOS and leads to a small Seebeck coefficient, while $Ca_{14}MgSb_{10.9}Sn_{0.1}$ and $Ca_{14}MgSb_{10.8}Sn_{0.2}$ have Fermi-levels with a large DOS and a large Seebeck coefficient. Mobility is inversely related to resistivity by the equation, $1/\rho = \sigma = ne\mu$ (n is carrier density, e is the charge carrier, u is mobility). Mobility depends on the band structure and affects resistivity significantly as the carrier densities for these samples are approximately the same.

5. Summary

$Ca_{14}MgSb_{11-x}Sn_x$ (x = 0.05, 0.1 and 0.2) solid solutions were synthesized by powder metallurgy with Mg_2Sn as precursor. Although the Sn amount in the samples is small (<2%), optimized Seebeck coefficient and resistivity were achieved simultaneously with similar thermal conductivity. This is very rare, as the Seebeck coefficient and resistivity are inversely related with respected to carrier concentration. The Seebeck coefficient of $Ca_{14}MgSb_{10.95}Sn_{0.05}$ has a similar temperature dependence compared with $Ca_{14}MgSb_{11}$, while the Seebeck coefficients of $Ca_{14}MgSb_{10.90}Sn_{0.10}$ and $Ca_{14}MgSb_{10.80}Sn_{0.20}$ linearly increase from room temperature to high temperature, different from the trend of $Ca_{14}MgSb_{11}$. Their resistivity shows semiconductor behavior and the activation energy decreases with Sn amount. The zT value of $Ca_{14}MgSb_{10.80}Sn_{0.20}$ reaches 0.49 at 1075 K, which is 53% higher than that of $Ca_{14}MgSb_{11}$ at the same temperature. The band structure of $Ca_{14}MgSb_7Sn_4$ is calculated to better understand the effect of Sn substitution. The band gap of $Ca_{14}MgSb_7Sn_4$ is 0.2 eV and the Fermi-level shifts into the valence band. $5p$-orbitals of Sn contribute most to the valence bands near Fermi-level at the top of valence band and the overall DOS of $Ca_{14}MgSb_7Sn_4$ are smaller in the valence band compared to $Ca_{14}MgSb_{11}$. The substitution of Sn increases the carrier concentration and decreases both the Seebeck coefficient and resistivity as expected.

Author Contributions: Y.H. and S.M.K conceived and designed the experiments. Y.H. performed the experiments and analyzed the data. K.A.L. contributed to the DOS calculations. Y.H. wrote the first draft of the manuscript; all co-authors contributed to the final version of the manuscript.

Acknowledgments: The authors thank Julia Zaikina for her help in TB-LMTO calculations. This work was supported by the NASA Science Mission Directorate's Radioisotope Power Systems. Financial support from NEUP is gratefully acknowledged.

Conflicts of Interest: The authors declare no conflict of interest.

References

1. Pomrehn, G.S.; Zevalkink, A.; Zeier, W.G.; van de Walle, A.; Snyder, G.J. Defect-Controlled Electronic Properties in AZn_2Sb_2 Zintl Phases. *Angew. Chem. In. Ed.* **2014**, *53*, 3422–3426. [CrossRef] [PubMed]
2. Grebenkemper, J.H.; Hu, Y.; Barrett, D.; Gogna, P.; Huang, C.-K.; Bux, S.K.; Kauzlarich, S.M. High Temperature Thermoelectric Properties of $Yb_{14}MnSb_{11}$ Prepared from Reaction of MnSb with the Elements. *Chem. Mater.* **2015**, *27*, 5791. [CrossRef]

3. Zevalkink, A.; Zeier, W.G.; Cheng, E.; Snyder, J.; Fleurial, J.-P.; Bux, S. Nonstoichiometry in the Zintl Phase $Yb_{1-\delta}Zn_2Sb_2$ as a Route to Thermoelectric Optimization. *Chem. Mater.* **2014**, *26*, 5710. [CrossRef]

4. Toberer, E.S.; Cox, C.A.; Brown, S.R.; Ikeda, T.; May, A.F.; Kauzlarich, S.M.; Snyder, G.J. Traversing the Metal-Insulator Transition in a Zintl Phase: Rational Enhancement of Thermoelectric Efficiency in $Yb_{14}Mn_{1-x}Al_xSb_{11}$. *Adv. Funct. Mater.* **2008**, *18*, 2795. [CrossRef]

5. Toberer, E.S.; Rauwel, P.; Gariel, S.; Taftø, J.; Jeffrey Snyder, G. Composition and the Thermoelectric Performance of β-Zn_4Sb_3. *J. Mater. Chem.* **2010**, *20*, 9877. [CrossRef]

6. Snyder, G.J.; Christensen, M.; Nishibori, E.; Caillat, T.; Iversen, B.B. Disordered Zinc in Zn_4Sb_3 with Phonon-glass and Electron-crystal Thermoelectric Properties. *Nat. Mater.* **2004**, *3*, 458. [CrossRef] [PubMed]

7. He, Y.; Zhang, T.; Shi, X.; Wei, S.H.; Chen, L. High Thermoelectric Performance in Copper Telluride. *NPG Asia Mater.* **2015**, *7*, e210. [CrossRef]

8. He, Y.; Day, T.; Zhang, T.; Liu, H.; Shi, X.; Chen, L.; Snyder, G.J. High Thermoelectric Performance in Non-Toxic Earth-abundant Copper Sulfide. *Adv. Mater.* **2014**, *26*, 3974. [CrossRef] [PubMed]

9. Ge, Z.H.; Zhang, B.P.; Chen, Y.X.; Yu, Z.X.; Liu, Y.; Li, J.F. Synthesis and Transport Property of $Cu_{1.8}S$ as a Promising Thermoelectric Compound. *Chem. Commun.* **2011**, *47*, 12697. [CrossRef] [PubMed]

10. Liu, H.; Shi, X.; Xu, F.; Zhang, L.; Zhang, W.; Chen, L.; Li, Q.; Uher, C.; Day, T.; Snyder, G.J. Copper Ion Liquid-like Thermoelectrics. *Nat. Mater.* **2012**, *11*, 422. [CrossRef] [PubMed]

11. Kaltzoglou, A.; Fässler, T.; Christensen, M.; Johnsen, S.; Iversen, B.; Presniakov, I.; Sobolev, A.; Shevelkov, A. Effects of the Order–disorder Phase Transition on the Physical Properties of $A_8Sn_{44}\square_2$ (A = Rb, Cs). *J. Mater. Chem.* **2008**, *18*, 5630. [CrossRef]

12. Kaltzoglou, A.; Fässler, T.F.; Gold, C.; Scheidt, E.W.; Scherer, W.; Kume, T.; Shimizu, H. Investigation of Substitution Effects and the Phase Transition in Type-I clathrates $Rb_xCs_{8-x}Sn_{44}\square_2$ ($1.3 \leq x \leq 2.1$) Using Single-crystal X-ray Diffraction, Raman Spectroscopy, Heat Capacity and Electrical Resistivity Measurements. *J. Solid State Chem.* **2009**, *182*, 2924. [CrossRef]

13. Chung, D.; Hogan, T.; Brazis, P.; Roccilane, M.; Kannewurf, C.; Bastea, M.; Uher, C.; Kanatzidis, M.G. $CsBi_4Te_6$: A High-Performance Thermoelectric Material for Low-Temperature Applications. *Science* **2000**, *287*, 1024. [CrossRef] [PubMed]

14. Dolyniuk, J.A.; Owens-Baird, B.; Wang, J.; Zaikina, J.V.; Kovnir, K. Clathrate Thermoelectrics. *Mater. Sci. Eng. R* **2016**, *108*, 1. [CrossRef]

15. Zaikina, J.V.; Kovnir, K.A.; Sobolev, A.V.; Presniakov, I.A.; Prots, Y.; Baitinger, M.; Schnelle, W.; Olenev, A.V.; Lebedev, O.I.; Van Tendeloo, G.; et al. $Sn_{20.5}\square_{3.5}As_{22}I_8$: A Largely Disordered Cationic Clathrate with a New Type of Superstructure and Abnormally Low Thermal Conductivity. *Chem. Eur. J.* **2007**, *13*, 5090. [CrossRef] [PubMed]

16. Aydemir, U.; Zevalkink, A.; Ormeci, A.; Gibbs, Z.M.; Bux, S.; Snyder, G.J. Thermoelectric Enhancement in $BaGa_2Sb_2$ by Zn Doping. *Chem. Mater.* **2015**, *27*, 1622. [CrossRef]

17. Chanakian, S.; Aydemir, U.; Zevalkink, A.; Gibbs, Z.M.; Fleurial, J.-P.; Bux, S.; Snyder, G.J. High Temperature Thermoelectric Properties of Zn-doped $Eu_5In_2Sb_6$. *J. Mater. Chem. C* **2015**, *3*, 10518. [CrossRef]

18. Chanakian, S.; Zevalkink, A.; Aydemir, U.; Gibbs, Z.M.; Pomrehn, G.; Fleurial, J.-P.; Bux, S.; Snyder, G.J. Enhanced Thermoelectric Properties of $Sr_5In_2Sb_6$ via Zn-doping. *J. Mater. Chem. A* **2015**, *3*, 10289. [CrossRef]

19. Heremans, J.P.; Jovovic, V.; Toberer, E.S.; Saramat, A.; Kurosaki, K.; Charoenphakdee, A.; Yamanaka, S.; Snyder, G.J. Enhancement of Thermoelectric Efficiency in PbTe by Distortion of the Electronic Density of States. *Science* **2008**, *321*, 554. [CrossRef] [PubMed]

20. Zhang, Q.; Cao, F.; Liu, W.; Lukas, K.; Yu, B.; Chen, S.; Opeil, C.; Broido, D.; Chen, G.; Ren, Z. Heavy Doping and Band Engineering by Potassium to Improve the Thermoelectric Figure of Merit in p-Type PbTe, PbSe, and $PbTe_{1-y}Se_y$. *J. Am. Chem. Soc.* **2012**, *134*, 10031. [CrossRef] [PubMed]

21. Pei, Y.; LaLonde, A.; Iwanaga, S.; Snyder, G.J. High Thermoelectric Figure of Merit in Heavy Hole Dominated PbTe. *Energy Environ. Sci.* **2011**, *4*, 2085. [CrossRef]

22. Girard, S.N.; He, J.; Zhou, X.; Shoemaker, D.; Jaworski, C.M.; Uher, C.; Dravid, V.P.; Heremans, J.P.; Kanatzidis, M.G. High Performance Na-doped PbTe-PbS Thermoelectric Materials: Electronic Density of States Modification and Shape-controlled Nanostructures. *J. Am. Chem. Soc.* **2011**, *133*, 16588. [CrossRef] [PubMed]

23. Zhao, L.D.; Tan, G.; Hao, S.; He, J.; Pei, Y.; Chi, H.; Wang, H.; Gong, S.; Xu, H.; Dravid, V.P.; et al. Ultrahigh Power Factor and Thermoelectric Performance in Hole-doped Single-Crystal SnSe. *Science* **2016**, *351*, 141. [CrossRef] [PubMed]

24. Wu, D.; Zhao, L.D.; Hao, S.; Jiang, Q.; Zheng, F.; Doak, J.W.; Wu, H.; Chi, H.; Gelbstein, Y.; Uher, C.; et al. Origin of the High Performance in GeTe-Based Thermoelectric Materials upon Bi_2Te_3 Doping. *J. Ame. Chem. Soc.* **2014**, *136*, 11412. [CrossRef] [PubMed]

25. Levin, E.M.; Cook, B.A.; Harringa, J.L.; Bud'ko, S.L.; Venkatasubramanian, R.; Schmidt-Rohr, K. Analysis of Ce- and Yb-Doped TAGS-85 Materials with Enhanced Thermoelectric Figure of Merit. *Adv. Funct. Mater.* **2011**, *21*, 441. [CrossRef]

26. Levin, E.M.; Bud'ko, S.L.; Schmidt-Rohr, K. Enhancement of Thermopower of TAGS-85 High-Performance Thermoelectric Material by Doping with the Rare Earth Dy. *Adv. Funct. Mater.* **2012**, *22*, 2766. [CrossRef]

27. Bhattacharya, S.; Pope, A.L.; Littleton, R.T.; Tritt, T.M.; Ponnambalam, V.; Xia, Y.; Poon, S.J. Effect of Sb Doping on the Thermoelectric Properties of Ti-based half-Heusler Compounds, $TiNiSn_{1-x}Sb_x$. *Appl. Phys. Lett.* **2000**, *77*, 2476. [CrossRef]

28. Shen, Q.; Chen, L.; Goto, T.; Hirai, T.; Yang, J.; Meisner, G.P.; Uher, C. Effects of Partial Substitution of Ni by Pd on the Thermoelectric Properties of ZrNiSn-based half-Heusler Compounds. *Appl. Phys. Lett.* **2001**, *79*, 4165. [CrossRef]

29. Appel, O.; Zilber, T.; Kalabukhov, S.; Beeri, O.; Gelbstein, Y. Morphological Effects on the Thermoelectric Properties of $Ti_{0.3}Zr_{0.35}Hf_{0.35}Ni1+\delta Sn$ Alloys Following Phase Separation. *J. Mater. Chem. C* **2015**, *3*, 11653. [CrossRef]

30. Casper, F.; Graf, T.; Chadov, S.; Balke, B.; Felser, C. Half-Heusler Compounds: Novel Materials for Energy and Spintronic Applications. *Semicond. Sci. Technol.* **2012**, *27*, 063001. [CrossRef]

31. Hu, Y.; Wang, J.; Kawamura, A.; Kovnir, K.; Kauzlarich, S.M. $Yb_{14}MgSb_{11}$ and $Ca_{14}MgSb_{11}$—New Mg-Containing Zintl Compounds and Their Structures, Bonding, and Thermoelectric Properties. *Chem. Mater.* **2015**, *27*, 343. [CrossRef]

32. Toberer, E.S.; Brown, S.R.; Ikeda, T.; Kauzlarich, S.M.; Jeffrey Snyder, G. High Thermoelectric Efficiency in Lanthanum Doped $Yb_{14}MnSb_{11}$. *Appl. Phys. Lett.* **2008**, *93*, 062110. [CrossRef]

33. Yi, T.; Abdusalyamova, M.N.; Makhmudov, F.; Kauzlarich, S.M. Magnetic and Transport Properties of Te Doped $Yb_{14}MnSb_{11}$. *J. Mater. Chem.* **2012**, *22*, 14378. [CrossRef]

34. Rauscher, J.F.; Cox, C.A.; Yi, T.; Beavers, C.M.; Klavins, P.; Toberer, E.S.; Snyder, G.J.; Kauzlarich, S.M. Synthesis, Structure, Magnetism, and High Temperature Thermoelectric Properties of Ge Doped $Yb_{14}MnSb_{11}$. *Dalton Trans.* **2010**, *39*, 1055. [CrossRef] [PubMed]

35. Rodríquez-Carvajal, J. Recent Advances in Magnetic Structure Determination by Neutron Powder Diffraction. *Physica B* **1993**, *192*, 55. [CrossRef]

36. Ravi, V.; Firdosy, S.; Caillat, T.; Brandon, E.; Van Der Walde, K.; Maricic, L.; Sayir, A. Thermal Expansion Studies of Selected High-Temperature Thermoelectric Materials. *J. Electron. Mater.* **2009**, *38*, 1433. [CrossRef]

37. Mackey, J.; Dynys, F.; Sehirlioglu, A. Uncertainty Analysis for Common Seebeck and Electrical Resistivity Measurement Systems. *Rev. Sci. Instrum.* **2014**, *85*, 085119. [CrossRef] [PubMed]

38. Andersen, O.K. Linear Methods in Band Theory. *Phys. Rev. B* **1975**, *12*, 3060. [CrossRef]

39. Andersen, O.K.; Jepsen, O. Explicit, First-Principles Tight-Binding Theory. *Phys. Rev. Lett.* **1984**, *53*, 2571. [CrossRef]

40. Andersen, O.K.; Pawlowska, Z.; Jepsen, O. Illustration of the Linear-muffin-tin-orbital Tight-binding Representation: Compact Orbitals and Charge Density in Si. *Phys. Rev. B* **1986**, *34*, 5253. [CrossRef]

41. Nowak, H.J.; Andersen, O.K.; Fujiwara, T.; Jepsen, O.; Vargas, P. Electronic-structure Calculations for Amorphous Solids Using the Recursion Method and Linear Muffin-tin Orbitals: Application to $Fe_{80}B_{20}$. *Phys. Rev. B* **1991**, *44*, 3577. [CrossRef]

42. Lambrecht, W.R.L.; Andersen, O.K. Minimal Basis Sets in the Linear Muffin-tin Orbital Method: Application to the Diamond-structure Crystals C, Si, and Ge. *Phys. Rev. B* **1986**, *34*, 2439. [CrossRef]

43. Brown, S.R.; Kauzlarich, S.M.; Gascoin, F.; Jeffrey Snyder, G. High-temperature Thermoelectric Studies of $A_{11}Sb_{10}$ (A=Yb, Ca). *J. Solid State Chem.* **2007**, *180*, 1414. [CrossRef]

44. Fistul, V.I. *Heavily Doped Semiconductor*; Plenum Press: New York, NY, USA, 1969.

45. Cutler, M.; Mott, N.F. Observation of Anderson Localization in an Electron Gas. *Phys. Rev.* **1969**, *181*, 1336. [CrossRef]

46. Mahan, G.D.; Sofo, J.O. The Best Thermoelectric. *Proc. Natl. Acad. Sci. USA* **1996**, *93*, 7436. [CrossRef] [PubMed]

crystals

MDPI

Article

Lu$_5$Pd$_4$Ge$_8$ and Lu$_3$Pd$_4$Ge$_4$: Two More Germanides among Polar Intermetallics

Riccardo Freccero [1], Pavlo Solokha [1,*], Davide Maria Proserpio [2,3], Adriana Saccone [1] and Serena De Negri [1]

[1] Dipartimento di Chimica e Chimica Industriale, Università degli Studi di Genova, Via Dodecaneso 31, 16146 Genova, Italy; riccardo.freccero@edu.unige.it (R.F.); adriana.saccone@unige.it (A.S.); serena.denegri@unige.it (S.D.N.)

[2] Dipartimento di Chimica, Università degli Studi di Milano, Via Golgi 19, 20133 Milano, Italy; davide.proserpio@unimi.it

[3] Samara Center for Theoretical Materials Science (SCTMS), Samara State Technical University, Molodogvardeyskaya St. 244, Samara 443100, Russia

*   Correspondence: pavlo.solokha@unige.it; Tel.: +39-010-3536-159

Received: 20 April 2018; Accepted: 3 May 2018; Published: 5 May 2018

Abstract: In this study, two novel Lu$_5$Pd$_4$Ge$_8$ and Lu$_3$Pd$_4$Ge$_4$ polar intermetallics were prepared by direct synthesis of pure constituents. Their crystal structures were determined by single crystal X-ray diffraction analysis: Lu$_5$Pd$_4$Ge$_8$ is monoclinic, $P2_1/m$, $mP34$, a = 5.7406(3), b = 13.7087(7), c = 8.3423(4) Å, β = 107.8(1), Z = 2; Lu$_3$Pd$_4$Ge$_4$ is orthorhombic, $Immm$, $oI22$, a = 4.1368(3), b = 6.9192(5), c = 13.8229(9) Å, Z = 2. The Lu$_5$Pd$_4$Ge$_8$ analysed crystal is one more example of non-merohedral twinning among the rare earth containing germanides. Chemical bonding DFT studies were conducted for these polar intermetallics with a metallic-like behavior. Gathered results for Lu$_5$Pd$_4$Ge$_8$ and Lu$_3$Pd$_4$Ge$_4$ permit to described both of them as composed by [Pd–Ge]$^{8-}$ three dimensional networks bonded to positively charged lutetium species. From the structural chemical point of view, the studied compounds manifest some similarities to the Zintl phases, containing well-known covalent fragments i.e., Ge dumbbells as well as unique *cis*-Ge$_4$ units. A comparative analysis of molecular orbital diagrams for Ge$_2^{6-}$ and *cis*-Ge^{10-} anions with COHP results supports the idea of the existence of complex Pd–Ge polyanions hosting covalently bonded partially polarised Ge units. The palladium atoms have an anion like behaviour and being the most electronegative cause the noticeable variation of Ge species charges from site to site. Lutetium charges oscillate around +1.5 for all crystallographic positions. Obtained results explained why the classical Zintl-Klemm concept can't be applied for the studied polar intermetallics.

Keywords: polar intermetallics; symmetry reduction; chemical bond

1. Introduction

In *RE*–Pd–Ge systems (*RE* = rare earth metal) more than one hundred ternary compounds have already been discovered [1], which have been extensively studied with respect to crystal structure, chemical bonding and physical properties [2–6].

The structures of Ge-rich compounds are characterized by a variety of Ge covalent fragments, with topologies depending both on global stoichiometry and on the nature of the *RE* component. These units are often joined together through Pd atoms, meanwhile the *RE* species are located in bigger channels inside the structure [2,3,7]. The frameworks formed by Pd and Ge atoms have been interpreted as polyanions of general formula [Pd$_x$Ge$_y$]$^{8-}$ counterbalanced by the rare earth cations, coherently with the definition of these compounds as polar intermetallics [4].

It is interesting to remark that the ternary *RE*–Pd–Ge compounds manifest a tendency to be stoichiometric with ordered distributions of constituents through distinct Wyckoff sites. Moreover, within Pd–Ge fragments, both species have small coordination numbers (usually four or five) with very similar topological distributions of neighbours (tetrahedral coordination or its derivatives). These features may be considered as geometrical traces of a similar chemical role of Pd and Ge. That is why symmetry reduction from certain aristotypes can conveniently depict the distortions related with an ordered distribution of atom sorts. Such analysis has been conducted in the literature for AlB$_2$ derivative polymorphs of *RE*PdGe [8] and BaAl$_4$ derivatives of the *RE*$_2$Pd$_3$Ge$_5$ [7,9] family of compounds. In systems where such types of relationships exist, the geometric factor is surely of great importance. Thus, varying *RE*, different polymorphs [8] or even novel compounds may form. As an example, heavy rare earth containing *RE*$_5$Pd$_4$Ge$_8$ (*RE* = Er, Tm) [4] and *RE*$_3$Pd$_4$Ge$_4$ (*RE* = Ho, Tm, Yb) [3] series of compounds may be cited.

During exploratory syntheses conducted in the Lu–Pd–Ge system in the framework of our ongoing studies on Ge-rich ternary compounds, the Lu representatives of the abovementioned 5:4:8 and 3:4:4 stoichiometries were detected for the first time. In this paper, results on the synthesis and structural characterization/analysis of these new germanides are reported, together with an extensive study of their chemical bonding, including Bader charges, Density of States (DOS) and Crystal Orbital Hamilton Population (COHP) curves as well as Molecular Orbitals (MO) diagrams for Zintl anions composed by Ge.

2. Experimental

2.1. Synthesis and SEM-EDXS Characterization

The Lu–Pd–Ge alloys were synthesized from elements with nominal purities >99.9% mass. Lutetium was supplied by Newmet Koch, Waltham Abbey, England, and palladium and germanium by MaTecK, Jülich, Germany.

Different synthetic routes were followed, including arc melting and direct synthesis in resistance furnace. In the latter case, proper amounts of components were placed in an alumina crucible, which was closed in an evacuated quartz ampoule to prevent oxidation at high temperatures, and submitted to one of the following thermal cycles in a resistance furnace:

(1) 25 °C → (10 °C/min) → 950 °C (1 h) → (−0.2 °C/min) → 600 °C (168 h) → (−0.5 °C/min) → 300 °C → furnace switched off

(2) 25 °C → (10 °C/min) → 1150 °C (1 h) → (−0.2 °C/min) → 300 °C → furnace switched off

A continuous rotation of the quartz ampoule during the thermal cycle was applied. In some cases, the thermal treatment followed arc melting. A scanning electron microscope (SEM) Zeiss Evo 40 (Carl Zeiss SMT Ltd., Cambridge, UK) coupled with a Pentafet Link Energy Dispersive X-ray Spectroscopy (EDXS) system managed by INCA Energy software (Oxford Instruments, Analytical Ltd., Bucks, UK) was used for microstructure observation and phase analysis. For this last purpose, calibration was performed with a cobalt standard. Samples to be analyzed were embedded in a phenolic resin with carbon filler, by using the automatic hot compression mounting press, Opal 410 (ATM GmbH, Mammelzen, Germany), and smooth surfaces for microscopic examinations were obtained with the aid of the automatic grinding and polishing machine, Saphir 520 (ATM GmbH, Mammelzen, Germany). SiC papers with grain sizes decreasing from 600 to 1200 mesh and diamond pastes with particle sizes decreasing from 6 to 1 μm were employed for grinding and polishing, respectively.

2.2. X-ray Diffraction (XRD) Measurements on Single Crystals and Powder Samples

Single crystals of Lu$_5$Pd$_4$Ge$_8$ and Lu$_3$Pd$_4$Ge$_4$ were selected from suitable samples with the aid of a light optical microscope operated in dark field mode. A full-sphere dataset was obtained in a routine fashion at ambient conditions on a four-circle Bruker Kappa APEXII CCD area-detector diffractometer equipped by the graphite monochromatized Mo $K\alpha$ (λ = 0.71073 Å) radiation, operating in ω-scan mode. Crystals exhibiting metallic luster and glued on glass fibers were mounted in a goniometric head and then placed in a goniostat inside a diffractometer camera. Intensity data were collected over the reciprocal space up to ~30° in θ with exposures of 20 s per frame. Semi-empirical absorption corrections based on a multipolar spherical harmonic expansion of equivalent intensities were applied to all data by the SADABS/TWINABS (2008) software [10].

The corresponding CIF files are available in the supporting information material and they have also been deposited at Fachinformationszentrum Karlsruhe, 76344 Eggenstein-Leopoldshafen, Germany, with the following depository numbers: CSD-434226 ($Lu_5Pd_4Ge_8$) and CSD-434225 ($Lu_3Pd_4Ge_4$). Selected crystallographic data and structure refinement parameters for the studied single crystals are listed in Table 1. Details regarding the structure solution are discussed in Section 3.2. X-ray powder diffraction (XRPD) measurements were performed on all samples, using a Philips *X'Pert* MPD vertical diffractometer (Cu $K\alpha$ radiation, λ = 1.5406 Å, graphite crystal monochromator, scintillation detector, step mode of scanning). Phase identification was performed with the help of the PowderCell software, version 2.4 [11].

Table 1. Crystallographic data for $Lu_5Pd_4Ge_8$ and $Lu_3Pd_4Ge_4$ single crystals together with some experimental details of their structure determination.

Empirical Formula	$Lu_5Pd_4Ge_8$	$Lu_3Pd_4Ge_4$
EDXS data	$Lu_{28.6}Pd_{24.9}Ge_{46.5}$	$Lu_{25.7}Pd_{35.0}Ge_{39.5}$
Space group (No.)	$P2_1/m$ (11)	*Immm* (71)
Pearson symbol-prototype, Z	$mP34$-$Tm_5Pd_4Ge_8$, 2	$oI22$-$Gd_3Cu_4Ge_4$, 2
a [Å]	5.7406(3)	4.1368(3)
b [Å]	13.7087(7)	6.9192(5)
c [Å]	8.3423(4)	13.8229(9)
β (°)	107.8(1)	–
V [Å^3]	625.20(5)	395.66(5)
Abs. coeff. (μ), mm^{-1}	63.5	60.7
Twin law	$[-\frac{1}{2}\,0\,\frac{1}{2}; 0-1\,0;\,\frac{3}{2}\,0\,\frac{1}{2}]$	–
k (BASF)	0.49(1)	–
Unique reflections	2105	404
Reflections I > 2σ(I)/parameters	1877/87	398/23
GOF on F^2 (S)	1.17	1.17
R indices [I > 2σ(I)]	R1 = 0.0190;wR2 = 0.0371	R1 = 0.0238;wR2 = 0.0869
R indices [all data]	R1 = 0.0247;wR2 = 0.0384	R1 = 0.0242;wR2 = 0.0871
$\Delta\varrho_{fin}$ (max/min), [e/Å^3]	2.00/−2.83	2.87/−3.33

2.3. Computational Details

A charge analysis based on Bader's Quantum Theory of Atoms In Molecules (QTAIM) [12], coded in the Vienna Ab-initio Simulation Package (VASP) [13], was used to evaluate the atomic charge populations in the title compounds. Projector augmented waves (PAW) formalism was used, together with Perdew–Berke–Erzenhof parametrization of the exchange-correlation interaction. The recommended PAW sets were used, considering nine valence electrons for Lu ($6s^25p^65d^1$), ten for Pd ($5s^14d^9$), and fourteen for Ge ($4s^23d^{10}4p^2$). An energy cut-off of 600 eV was set for all calculations presented and the default value (10^{-5} eV) of the energy convergence was used.

The electronic band structures of $Lu_5Pd_4Ge_8$ and $Lu_3Pd_4Ge_4$ were calculated by means of the self-consistent, tight-binding, linear-muffin-tin-orbital, atomic-spheres approximation method using the Stuttgart TB-LMTO-ASA 4.7 program [14], within the local density approximation (LDA) [15] of DFT. The radii of the Wigner–Seitz spheres were assigned automatically so that the overlapping potentials would be the best possible approximations to the full potential, and no empty spheres were needed to meet the minimum overlapping criterion.

The basis sets included $6s/(6p)/5d$ orbitals for Lu with Lu $4f^{14}$ treated as core, $5s/5p/4d/(4f)$ for palladium and $4s/4p/(4d)/(4f)$ orbitals for germanium with orbitals in parentheses being downfolded.

The Brillouin zone integrations were performed by an improved tetrahedron method using a $20 \times 8 \times 12$ k-mesh for $Lu_5Pd_4Ge_8$ and $16 \times 16 \times 16$ for $Lu_3Pd_4Ge_4$.

Crystal Orbital Hamilton populations (COHPs) [16] were used to analyze chemical bonding. The integrated COHP values (iCOHPs) were calculated in order to evaluate the strengths of different interactions. Plots of DOS and COHP curves were generated using wxDragon [17], setting the Fermi energy at 0 eV as a reference point.

Qualitative MO arguments based on extended Hückel theory (EHT) have been developed with the CACAO package [18,19] and its graphic interface. Even if the EHT model tends to involve the most drastic approximations in MO theory, this one electron effective Hamiltonian method tends to be used to generate qualitatively correct molecular and crystal orbitals [20]. EHT is best used to provide models for understanding both molecular and solid state chemistry, as shown with great success by Roald Hoffmann and others [21].

3. Results and Discussion

3.1. Results of SEM-EDXS Characterization

An explorative study of the Ge-rich region of the Lu–Pd–Ge system was conducted by synthesis of some ternary samples with a Ge content >40 at %. The prepared samples are listed in Table 2, together with an indication of the followed synthetic route, as well as the results of SEM/EDXS characterization. Information on phase crystal structure was obtained from X-ray diffraction results.

Table 2. Results of SEM/EDXS characterization of the Lu–Pd–Ge samples (> 40 at % Ge) obtained with different synthesis methods/thermal treatments. The highest yield phase in each sample is the first in the list.

No. Overall Composition [at %] Synthesis/Thermal Treatment	Phases	Phase Composition [at %] Lu; Pd; Ge	Crystal Structure
1 $Lu_{21.4}Pd_{11.2}Ge_{67.4}$ Arc melting followed by thermal treatment (1)	Lu_2PdGe_6 $Lu_5Pd_4Ge_8$ $LuPd_{0.16}Ge_2$ Ge	21.5; 12.1; 66.4 28.6; 25.1; 46.3 31.1; 5.4; 63.5 –; –; –;	$oS72–Ce_2(Ga_{0.1}Ge_{0.9})_7$ $mP34-Tm_5Pd_4Ge_8$ $oS16-CeNiSi_2$ $cF8-C$
2 $Lu_{28.9}Pd_{24.1}Ge_{47.0}$ Arc melting	$Lu_5Pd_4Ge_8$ new phase Lu_2PdGe_6 $LuPd_{0.16}Ge_2$	28.8; 24.8; 46.4 32.4; 28.5; 39.1 21.7; 11.8; 66.5 30.1; 6.9; 63.0	$mP34-Tm_5Pd_4Ge_8$ AlB_2 related $oS72–Ce_2(Ga_{0.1}Ge_{0.9})_7$ $oS16-CeNiSi_2$
3 * $Lu_{30.8}Pd_{25.5}Ge_{43.7}$ Direct synthesis with thermal treatment (2)	$Lu_5Pd_4Ge_8$ new phase Ge	28.6; 24.9; 46.5 33.0; 26.8; 40.2 –; –; –;	$mP34-Tm_5Pd_4Ge_8$ AlB_2 related $cF8-C$
4 * $Lu_{33.0}Pd_{26.0}Ge_{41.0}$ Arc melting followed by thermal treatment (2)	$Lu_3Pd_4Ge_4$ $Lu_5Pd_4Ge_8$ LuPdGe PdGe Ge	25.7; 35.0; 39.5 28.4; 25.1; 46.5 31.9; 34.5; 33.6 0.0; 53.4; 47.6 0.0; 0.0; 100.0	$oI22-Gd_3Cu_4Ge_4$ $mP34-Tm_5Pd_4Ge_8$ $oI36-AuYbSn$ $oP8-FeAs$ $cF8-C$
5 $Lu_{17.9}Pd_{29.0}Ge_{53.1}$ Arc melting	$Lu_3Pd_4Ge_4$ LuPdGe PdGe Ge	26.1; 34.2; 39.7 32.0; 33.5; 34.5 0.0; 52.4; 47.8 0.0; 0.0; 100.0	$oI22-Gd_3Cu_4Ge_4$ $oI36-AuYbSn$ $oP8-FeAs$ $cF8-C$

* Samples from which single crystals were taken.

All samples are multiphase, as it is common for non-annealed alloys belonging to complex ternary systems, Ge is always present, in some cases in small amount. SEM images using the Back-Scattered Electron (BSE) mode are well contrasted, helping to distinguish different compounds, whose compositions are highly reproducible.

Several ternary compounds already known from the literature were detected in the samples, namely Lu_2PdGe_6, $LuPd_{0.16}Ge_2$ and LuPdGe [1,2]. For the latter, the *oI*36-AuYbSn structure was confirmed, in agreement with previous single crystal data [8].

A new phase of composition ~$Lu_{33}Pd_{27}Ge_{40}$ was detected in samples 2 and 3; the corresponding X-ray powder patterns could be acceptably indexed assuming a simple AlB_2-like structure, with a ≈ 4.28 and c ≈ 3.54 Å. Nevertheless, a deeper structural investigation would be necessary to ensure its crystal structure.

Crystal structures of the new $Lu_5Pd_4Ge_8$ and $Lu_3Pd_4Ge_4$ compounds were solved by analysing single crystals extracted from samples 3 and 4, respectively. The obtained structural models, discussed in the following section, were consistent with the measured powder patterns.

3.2. Crystal Structures of $Lu_5Pd_4Ge_8$ and $Lu_3Pd_4Ge_4$

3.2.1. Structural Determination

The $Lu_5Pd_4Ge_8$ crystal selected for X-ray analysis is one more example of non-merohedral twins among germanides. Previously, similar twins were found for Tb_3Ge_5 [22], Eu_3Ge_5 [23], Pr_4Ge_7 [24], La_2PdGe_6 and Pr_2PdGe_6 [2]. Based on the preliminary indexing results, the unit cell of the measured crystal might be considered as a base centered orthorhombic one with a = 8.55, b = 21.29 and c = 13.70 Å. The analysis of systematic extinctions suggested the following space groups: $Cmc2_1$ (No. 36), C2cm (No. 40) and Cmcm (No. 63). It should be mentioned that the average value of $|E^2–1|$ = 1.33, characterizing the distribution of peak intensities, deviates noticeably from the ideal value (0.968) for centrosymmetric space groups. Frequently, this is an indication of a twinned dataset [25,26]. A charge-flipping algorithm implemented in JANA2006 [27] was used, giving a preliminary structural model with 36 Lu atoms and 96 Ge atoms in the unit cell (Cmcm space group). Usually, when scatterers have such remarkable differences in electrons, the charge-flipping algorithm is quick and very efficient in discriminating them. Considering the interatomic distances criterion and U_{eq} values, in the successive iteration cycles, Pd atoms were introduced manually by substituting those of Ge, but no improvements were observed. There was no chance to improve this model further because the isotropic thermal displacement parameters showed meaningless values; several additional strong peaks were present at difference Fourier maps located too close to the accepted atom positions; and the R1 value stuck at ca. 10%. Looking for a correct structure solution in other space groups gave no reasonable results.

At this point, a more careful analysis of diffraction spots in reciprocal space was performed using RLATT [10] software. It was noticed that a remarkable number of peaks distributed in a regular way had a small intensity and might be considered as super reflections. Therefore, they were ignored during the indexing procedure, and a four times smaller primitive monoclinic unit cell with a = 5.73, b= 13.70, c = 8.34 Å and β = 107.8° was derived. The dataset was newly integrated and semi-empirical absorption corrections were applied by SADABS [10] software. This time, an *mP*34 structural model, containing all the atomic species, was proposed by the charge-flipping algorithm. Even so, the refinement was not satisfactory because some Wyckoff sites manifested partial occupancy and it was not possible to refine the structure anisotropically. It was decided to test the ROTAX [28] algorithm implemented in WinGx [29] and check the possibility of interpreting our crystal as a non-merohedral twin. In fact,

a two-fold rotation along the [101] direction $\begin{pmatrix} \frac{-1}{2} & 0 & \frac{1}{2} \\ 0 & -1 & 0 \\ \frac{3}{2} & 0 & \frac{1}{2} \end{pmatrix}$ was proposed as a twin law obtaining

a good figure of merit.

To check this hypothesis and refine the collected data as accurately as possible, the initially selected batch of ca. 1000 reflections (comprising those of weak intensity considered as super reflections) was separated into two groups with the help of the CELL_NOW [10] program, suggesting the same twin law for the two monoclinic domains. Successively, the information on the reciprocal domain orientation stored in the .p4p file was used to integrate the dataset considering the simultaneous presence of both domains. After that, the resulting intensities set was scaled, corrected for absorption and merged with the help of the TWINABS [10] program. As a result, the output in HKLF5 format with a flag indicating the original domains, was generated. Using the latter and testing one more time the charge flipping procedure, the structural model was immediately found and element species were correctly assigned. The $Lu_5Pd_4Ge_8$ was of monoclinic symmetry (space group $P2_1/m$, $mP34$-$Tm_5Pd_4Ge_8$) and contained 3 Lu, 2 Pd and 6 Ge crystallographic sites. All the atom positions were completely occupied and did not manifest any considerable amount of statistical mixture. The anisotropically refined $Lu_5Pd_4Ge_8$ showed excellent residuals and flat difference Fourier maps (see Table 1). The refined volume ratio of twinned domains was 0.49/0.51.

The RLATT program was used to generate a picture showing the distribution of X-ray diffraction spots originating from the two domains, differentiated by color, in Figure 1 (upper part). The distribution of the non-overlapped peaks of the second domain was also easily visible on the precession photo of the $h3l$ zone, demonstrated in Figure 1 (lower part). In the same figure, a schematic real space representation of the mutual orientation of the twinned-crystal components is shown.

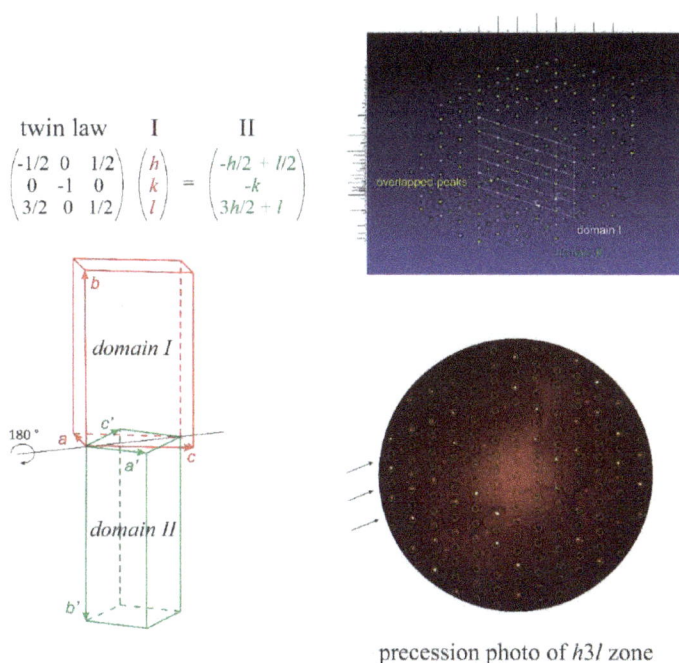

$$\text{twin law} \quad \text{I} \qquad \text{II}$$

$$\begin{pmatrix} -1/2 & 0 & 1/2 \\ 0 & -1 & 0 \\ 3/2 & 0 & 1/2 \end{pmatrix} \begin{pmatrix} h \\ k \\ l \end{pmatrix} = \begin{pmatrix} -h/2 + l/2 \\ -k \\ 3h/2 + l \end{pmatrix}$$

precession photo of $h3l$ zone

Figure 1. Twin law and reciprocal orientation of the two domains in the $Lu_5Pd_4Ge_8$ twinned crystal (**left**); distribution of the diffraction peaks in the reciprocal space (**right**). Nodes of the reciprocal pattern for each domain are shown in white and green, and overlapped peaks are yellow. On the experimental precession photos of the $h3l$ zone, arrows indicate the directions along which the second domain peaks are easily visible.

Indexing of the diffraction dataset of the $Lu_3Pd_4Ge_4$ single crystal gave an orthorhombic base centered unit cell with a = 4.137, b = 6.919, c = 13.823 Å. Systematic extinction conditions related to the presence of symmetry elements were not found for this dataset. The structure solution was found in *Immm* with the aid of the charge flipping algorithm implemented in JANA2006 [27]. The proposed preliminary structural model contained five crystallographic sites, giving the $Lu_3Pd_4Ge_4$ formula and corresponding to the *o*I22-$Gd_3Cu_4Ge_4$ prototype. Partial site occupation (due to a possible statistical mixture of the species) was checked in separate cycles of least-squares refinement, but no significant deviation from full occupation was detected. The final structure model was refined as stoichiometric with the anisotropic displacement parameters for all crystallographic sites, giving small residual factors and a flat difference Fourier map (see Table 1). The standardized atomic coordinates for $Lu_5Pd_4Ge_8$ and $Lu_3Pd_4Ge_4$ are given in Table 3.

Table 3. Atomic coordinates standardized by Structure Tidy [30] and equivalent isotropic displacement parameters for $Lu_5Pd_4Ge_8$ and $Lu_3Pd_4Ge_4$.

Atom	Site	x/a	y/b	z/c	U_{eq} (Å^2)
			$Lu_5Pd_4Ge_8$		
Lu1	2e	0.71858(8)	1/4	0.93028(6)	0.0047(1)
Lu2	4f	0.13606(7)	0.11370(2)	0.78913(7)	0.0051(1)
Lu3	4f	0.62176(8)	0.11902(2)	0.28943(7)	0.0056(1)
Pd1	4f	0.07436(13)	0.08476(3)	0.14089(12)	0.0072(1)
Pd2	4f	0.42601(13)	0.58211(3)	0.35985(12)	0.0075(1)
Ge1	2e	0.0515(2)	1/4	0.28977(15)	0.0081(2)
Ge2	2e	0.3343(2)	1/4	0.58221(15)	0.0078(2)
Ge3	2e	0.7797(2)	1/4	0.5814(2)	0.0063(2)
Ge4	4f	0.15453(17)	0.04252(5)	0.44776(16)	0.0071(1)
Ge5	2e	0.2797(2)	1/4	0.0606(2)	0.0048(2)
Ge6	4f	0.34622(17)	0.54443(4)	0.05049(16)	0.0060(1)
			$Lu_3Pd_4Ge_4$		
Lu1	2a	0	0	0	0.0110(2)
Lu2	4j	1/2	0	0.37347(4)	0.0081(2)
Pd	8l	0	0.30094(10)	0.32738(5)	0.0155(3)
Ge1	4h	0	0.18745(17)	1/2	0.0084(3)
Ge2	4i	0	0	0.21754(10)	0.0132(3)

Similar to $(Tm/Er)_5Pd_4Ge_8$ [4], the presence of Ge covalent fragments in $Lu_5Pd_4Ge_8$ is obvious. Among these, there were two almost identical Ge–Ge dumbbells distanced at 2.49 Å and one more finite fragment composed of four germanium atoms having a *cis*-configuration (Figure 2, Table 4). The latter manifests a small geometrical distortion from the ideal conformation due to slightly different chemical arrangements around terminal Ge atoms (terminal atoms are located at 2.56 and 2.63 Å far from central dumbbell; the internal obtuse angles are ca. 111° and 113°, respectively). The *cis* unit is planar and lays at the mirror plane of the $P2_1/m$ space group. The cited covalent fragments are joined together through Pd–Ge contacts shortened with respect to metallic radii sum (ranging from 2.51 to 2.73 Å) in a complex network hosting Lu atoms in the biggest cavities (see Figure 2). The shortest Lu–Pd and Lu–Ge contacts do not manifest noticeable deviations from the expected values and are ca. 3.0 Å.

Table 4. Interatomic distances and integrated crystal orbital Hamilton populations (-iCOHP, eV/cell) at E_F for the strongest contacts within the first coordination spheres in $Lu_5Pd_4Ge_8$. Symbols (2b) and (1b) indicate the number of homocontacts for corresponding Ge species.

Central Atom	Adjacent Atoms	d (Å)	-iCOHP	Central Atom	Adjacent Atoms	d (Å)	-iCOHP	Central Atom	Adjacent Atoms	d (Å)	-iCOHP
Lu1	Ge6 (×2)	2.853	1.26	Lu3	Ge5	2.904	1.25	(1b)Ge6	Ge6	2.494	2.39
	Ge1	3.025	0.81		Ge3	2.938	1.25		Pd1	2.516	2.16
	Ge5	3.033	1.02		Pd1	3.042	0.71		Pd2	2.533	2.12
	Ge3	3.036	0.85		Ge1	3.051	0.92		Pd1	2.619	1.80
	Ge2	3.064	0.79		Pd2	3.063	0.68		Lu1	2.853	1.26
	Ge5	3.069	0.81		Ge6	3.072	0.90		Lu2	3.016	0.96
	Pd1 (×2)	3.194	0.52		Ge6	3.094	0.75		Lu2	3.050	0.80
	Pd2 (×2)	3.258	0.46		Pd2	3.100	0.63		Lu3	3.072	0.90
Lu2	Ge5	2.857	1.29		Ge4	3.105	0.71		Lu3	3.094	0.75
	Ge3	2.918	1.20	(1b)Ge3	Ge4	3.120	0.84	Pd1	Ge6	2.516	2.16
	Ge2	2.994	0.99		Pd1	3.236	0.52		Ge4	2.526	2.11
	Ge6	3.016	0.95		Ge4	3.493	0.27		Ge1	2.606	1.66
	Ge4	3.043	0.91		Ge2	2.559	1.92		Ge6	2.619	1.80
	Ge4	3.043	0.79	(1b)Ge4	Pd2 (×2)	2.699	1.46		Ge5	2.730	1.35
	Ge6	3.050	0.80		Lu2 (×2)	2.918	1.20		Lu3	3.042	0.71
	Pd1	3.087	0.68		Lu3 (×2)	2.938	1.25		Lu2	3.087	0.68
	Pd1	3.104	0.63		Lu1	3.036	0.86		Lu2	3.104	0.63
	Pd2	3.114	0.64		Ge4	2.492	2.48		Lu1	3.194	0.52
	Pd2	3.156	0.57		Pd2	2.512	2.14	Pd2	Ge4	2.512	2.14
(2b)Ge1	Ge2	2.484	2.98		Pd1	2.526	2.11		Ge6	2.533	2.12
	Pd1 (×2)	2.606	1.66	(1b)Ge5	Pd2	2.566	1.94		Ge4	2.566	1.94
	Ge5	2.627	1.69		Lu2	3.043	0.91		Ge2	2.649	1.53
	Lu1	3.025	0.81		Lu2	3.043	0.79		Ge3	2.699	1.46
	Lu3 (×2)	3.051	0.92		Lu3	3.105	0.71		Lu3	3.063	0.68
(2b)Ge2	Ge1	2.484	2.98		Ge1	2.627	1.69		Lu3	3.100	0.63
	Ge3	2.559	1.92		Pd1 (×2)	2.730	1.35		Lu2	3.114	0.65
	Pd2 (×2)	2.649	1.53		Lu2 (×2)	2.857	1.29		Lu2	3.156	0.58
	Lu2 (×2)	2.994	1.00		Lu3 (×2)	2.904	1.26				
	Lu1	3.064	0.79		Lu1	3.033	1.02				
					Lu1	3.069	0.81				

The $Lu_3Pd_4Ge_4$ contains less germanium with respect to $Lu_5Pd_4Ge_8$ and, consequently, only a simple Ge–Ge dumbbell forms being, however, more stretched (2.59 Å, Table 5). The trend of other interactions is similar as for $Lu_5Pd_4Ge_8$; Pd and Ge construct an extended network with infinite channels of hexagonal and pentagonal forms hosting Lu atoms.

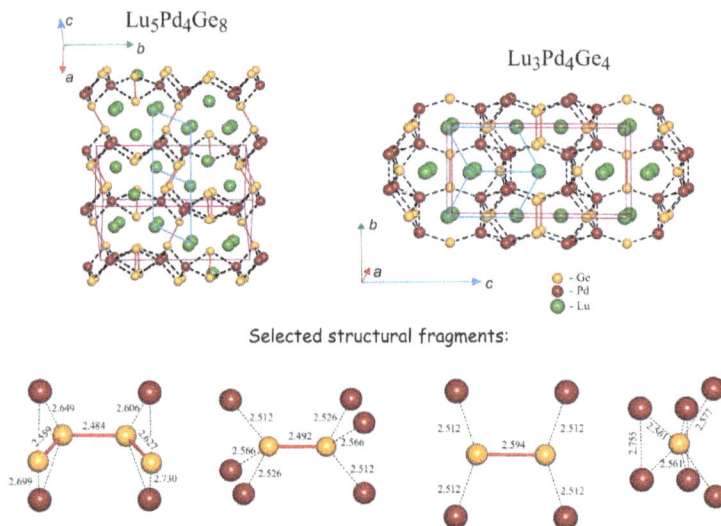

Figure 2. Crystal structures of $Lu_5Pd_4Ge_8$ and $Lu_3Pd_4Ge_4$. The Pd–Ge frameworks are evidenced by dotted lines. Ge–Ge covalent bonds are shown by red sticks. Selected fragments, discussed in the text, are pictured at the bottom. Selected interatomic distances (Å) are indicated. $ThCr_2Si_2$-like fragments are evidenced in blue.

Table 5. Interatomic distances and integrated crystal orbital Hamilton populations (-iCOHP, eV/cell) at E_F for the strongest contacts within the first coordination spheres in $Lu_3Pd_4Ge_4$. Symbols ($1b$) and ($0b$) indicate the number of homocontacts for corresponding Ge species.

Central Atom	Adjacent Atoms	d (Å)	-iCOHP	Central Atom	Adjacent Atoms	d (Å)	-iCOHP
Lu1	Ge4 (×4)	2.992	1.21	($0b$)Ge2	Pd (×4)	2.562	1.88
	Ge5 (×2)	3.006	1.05		Pd (×2)	2.577	1.86
	Pd (×8)	3.445	0.41		Lu2 (×2)	2.988	0.83
Lu2	Ge5 (×2)	2.988	0.83		Lu1	3.006	1.05
	Ge4 (×4)	3.003	0.99	Pd	Ge4	2.512	2.23
	Pd (×4)	3.003	0.79		Ge5 (×2)	2.562	1.88
	Pd (×2)	3.100	0.58		Pd	2.755	0.97
($1b$)Ge1	Pd (×2)	2.512	2.23		Lu2 (×2)	3.003	0.79
	Ge1	2.595	1.82		Pd (×2)	3.058	0.46
	Lu1 (×2)	2.992	1.22		Lu2	3.100	0.58
	Lu2 (×4)	3.003	0.99		Lu1 (×2)	3.445	0.41

One more structural relation can be proposed for the title compounds: both compounds contain common structural $ThCr_2Si_2$-like building blocks [31] (highlighted by blue lines in Figure 2) defined in many related compounds as "linkers" within various polyanionic fragments [32].

3.2.2. Lu$_5$Pd$_4$Ge$_8$: Structural Relationships

Looking for structural relationships is not an easy task, since this process is often strongly affected by human factors and is based on sometimes arbitrary criteria. From this point of view, one of the most rigorous approaches is based on the symmetry principle within the group-subgroup theory [33]. The most frequent chemical reason causing the reduction of symmetry is so-called "coloring", which can be interpreted as an ordered distribution of different chemical elements within distinct Wyckoff sites. Müller [34] and Pöttgen [35] depict numerous examples of these.

Structural relationships between Tm$_5$Pd$_4$Ge$_8$ (isostructural with Lu$_5$Pd$_4$Ge$_8$) and $RE_3T_2Ge_3$ (T = late transitional element) were proposed in the literature [2] based on topological similarities between polyanionic fragments and the spatial distribution of cations. An alternative description of relationships between the abovementioned structures in terms of symmetry reduction is proposed here. The stoichiometries of these compounds are related as follows:

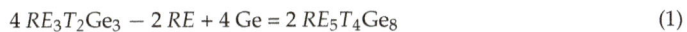

$$4\,RE_3T_2Ge_3 - 2\,RE + 4\,Ge = 2\,RE_5T_4Ge_8 \tag{1}$$

This relation, even if purely numerical, finds support when comparing the crystal structures of the two chemically affine representatives Lu$_3$Fe$_2$Ge$_3$ (*oS*32) and Lu$_5$Pd$_4$Ge$_8$ (*mP*34). As is evidenced in Figure 3, one of the Lu sites in the former is substituted by a Ge dumbbell in the latter.

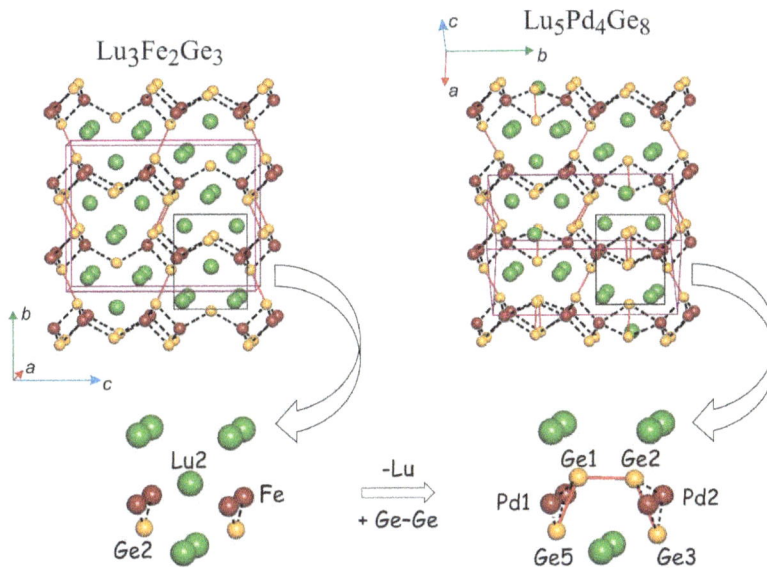

Figure 3. Structural similarities between Lu$_3$Fe$_2$Ge$_3$ and Lu$_5$Pd$_4$Ge$_8$. The polyanionic networks are shown by dotted lines, and covalent Ge fragments are joined by red sticks. The grey rectangle evidences regions of the crystal space where Lu/Ge$_2$ substitution takes place (for details see text).

From the chemical interaction point of view, this should be a drastic change; instead, the remaining atoms apparently do not suffer noticeable displacements. This is why it was checked whether a Bärnighausen tree might be constructed relating the *oS*32 and *mP*34 models. In fact, only two reduction steps were needed:

- a *traslationengleiche* (t2) decentering leading to a monoclinic Niggli cell (*mP*16-*P*2$_1$/*m*).
- a *klassengleiche* transformation (k2) giving a monoclinic model with doubled cell volume (*mP*32-*P*2$_1$/*m*). As a result, all the independent sites split in two (see Figure 4).

The Lu2′ site (2*e*: *0.211 1/4 0.430*) was further substituted by two germanium atoms (positions Ge1 and Ge2 in the final *mP34-P2$_1$/m* structural model). As a result, the already cited *cis*-Ge4 unit forms (see Figure 2), whose chemical role is discussed in the next section. The presence of the *cis*-Ge$_4$ units is quite intriguing, since the *trans* conformation is more favorable in numerous molecular chemistry examples. Therefore, it was decided to generate a structural model of Lu$_5$Pd$_4$Ge$_8$ composition hosting the *trans*-Ge unit and optimize it (see Figure S1 and Table S1). The relaxed structure perfectly coincided with the experimental results, confirming that minimal energy is associated with the *cis* conformation. More details on this, including an animation showing the evolution of the structural model after each relaxation step, are available in the Supplementary Material.

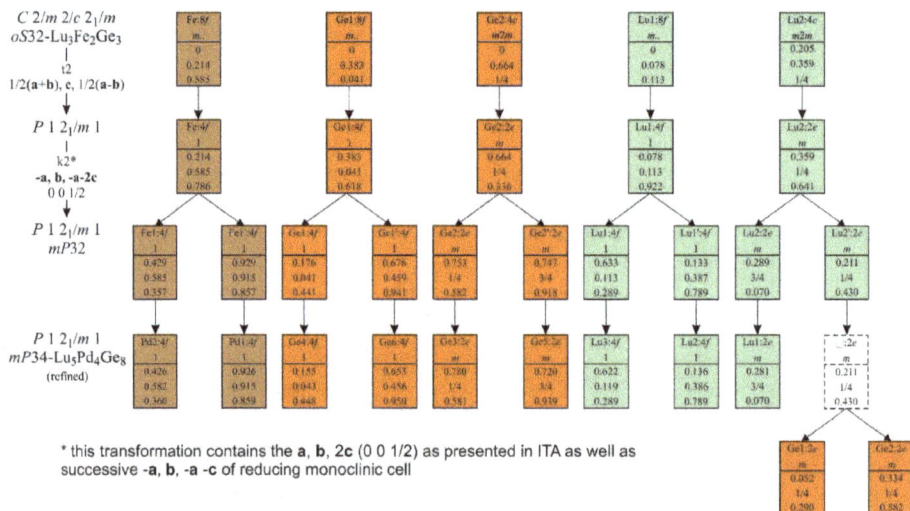

* this transformation contains the **a**, **b**, **2c** (0 0 1/2) as presented in ITA as well as successive -**a**, **b**, -**a** -**c** of reducing monoclinic cell

Figure 4. Evolution of the atomic parameters within the Bärnighausen formalism accompanying the symmetry reduction from Lu$_3$Fe$_2$Ge$_3$ to Lu$_5$Pd$_4$Ge$_8$ structures. The background colors correspond to the atom markers in the figures through the text.

3.3. Chemical Bonding Analysis

Frequently, chemical bonding in polar intermetallics is preliminary addressed using the *Zintl-Klemm* concept. Taking into account the interatomic distances between Ge atoms, the presence of [(1*b*)Ge^{3-}] with [(2*b*)Ge^{2-}] Zintl species in Lu$_5$Pd$_4$Ge$_8$ and [(1*b*)Ge^{3-}] with [(0*b*)Ge^{4-}] ones in Lu$_3$Pd$_4$Ge$_4$ could be guessed. In order to guarantee the precise electron count, the average number of valence electrons per Ge atom [*VEC*(Ge)] should amount to 6.75 for Lu$_5$Pd$_4$Ge$_8$ and to 7.50 for Lu$_3$Pd$_4$Ge$_4$. Although it is reasonable to hypothesize a formal charge transfer of 3 valence electrons per Lu atom (Lu^{3+}), as a first approximation, the Pd could be considered as a divalent cation (Pd^{2+}) or a neutral species (Pd^0). However, none of the possible electron distribution formulae listed below are suitable for the studied compounds, giving *VEC*(Ge) values that deviate somewhat from ideal values.

$$\text{Lu}_5\text{Pd}_4\text{Ge}_8 \ (\text{Pd}^0) \quad VEC(\text{Ge}) = 5.875$$

$$\text{Lu}_5\text{Pd}_4\text{Ge}_8 \ (\text{Pd}^{2+}) \quad VEC(\text{Ge}) = 6.875$$

$$\text{Lu}_3\text{Pd}_4\text{Ge}_4 \ (\text{Pd}^0) \quad VEC(\text{Ge}) = 6.250$$

$$\text{Lu}_3\text{Pd}_4\text{Ge}_4 \ (\text{Pd}^{2+}) \quad VEC(\text{Ge}) = 8.250$$

Even if the obtained *VEC*(Ge) values are closer to 6.75/7.50, in the case of Pd^{2+}, this assumption is not coherent with the valence electrons flow when considering any of the known electronegativity scales. For example, taking into account the *Pearson* electronegativity values for Pd (4.45 eV) and Ge (4.60 eV) it is clear that a charge transfer from Pd to Ge is hardly probable. Strictly speaking, it is not possible to successfully apply the (8–N) rule to interpret the Ge–Ge covalent interactions. Thus, it becomes clear that these simplified considerations are not sufficient to account for the chemical bonding of the studied intermetallics. In particular, it is not reliable to consider covalent Ge fragments as isolated and more complex interactions should be taken into account. Therefore, a deeper chemical bonding investigation was conducted.

In Table 6, the volumes of the atomic basins and Bader effective charges for all the atoms in $Lu_5Pd_4Ge_8$ and $Lu_3Pd_4Ge_4$ are listed together with those for the same species in their pure element form. Comparing these values, one can qualitatively estimate the chemical role of constituents in binary/ternary compounds.

Table 6. Calculated QTAIM effective charges and atomic basin volumes for Lu, Pd and Ge in their elemental structure, in $Lu_5Pd_4Ge_8$ and in $Lu_3Pd_4Ge_4$.

Element/Compound	Atom/Site	Volume, [Å³]	QTAIM Charge, Q^{eff}	Compound	Atom/Site	Volume, [Å³]	QTAIM Charge, Q^{eff}
Lu (*hP2*)	Lu/2*c*	29.74 #	0	$Lu_5Pd_4Ge_8$	Lu1/2*e*	15.88	+1.45
Pd (*cF4*)	Pd/4*a*	14.71 #	0	(*mP34*)	Lu2/4*f*	15.48	+1.48
Ge (*cF8*)	Ge/8*a*	22.66 #	0		Lu3/4*f*	15.85	+1.51
					Pd1/4*f*	19.91	−0.79
$Lu_3Pd_4Ge_4$	Lu1/2*a*	16.90	+1.57		Pd2/4*f*	19.77	−0.76
(*oI22*)	Lu2/4*j*	15.05	+1.53		(2*b*)Ge1/2*e*	19.39	−0.23
	Pd/8*l*	19.43	−0.67		(2*b*)Ge2/2*e*	19.51	−0.30
	(1*b*)Ge1/4*h*	22.35	−0.89		(1*b*)Ge3/2*e*	22.66	−0.87
	(0*b*)Ge2/4*i*	16.42	−0.09		(1*b*)Ge4/4*f*	18.73	−0.30
					(1*b*)Ge5/2*e*	23.96	−1.14
					(1*b*)Ge6/4*f*	19.67	−0.59

#—the QTAIM volumes of atoms in pure elements are equal to the volumes of their Wigner–Seitz polyhedra; structural data were taken from Ref [1].

In both ternary germanides, the QTAIM basins of Lu were shrunk with respect to Lu-*hP2*, and the corresponding charges oscillated around +1.5, confirming the active metal-like role of Lu. The significant difference between Lu effective charges and the formal charges suggest that some of its valence electrons may contribute to covalent interactions.

The palladium atoms had similar volumes of atomic basins (ca. 20 Å³) and are negatively charged (−0.7 ÷ −0.8), suggesting a bonding scenario coherent with the electronegativity values, i.e., with Pd taking part in a polyanionic network, as was hypothesized from the crystal structure analysis.

It is noteworthy that in the same compound, Ge atoms had pronounced differences in charge values (always negative) from site to site. More on the structural/chemical reasons for this will be discussed in the following.

The total and projected DOS for Lu, Pd and Ge for the studied intermetallics are shown in Figure 5. Orbital projected DOS can be found in the Supplementary Material (Figure S3). Focusing on the total DOS, a difference between the two compounds at the Fermi energy (E_F) is evident: for $Lu_5Pd_4Ge_8$ a pseudo-gap is visible just above E_F, instead for $Lu_3Pd_4Ge_4$ the Fermi level corresponds to a local maximum of the DOS, indicating a potential electronic instability. This might be a sign of particular physical properties (e.g., superconductivity or magnetic ordering) [36] or of small structural adjustments (e.g., off-stoichiometry due to statistical mixture or increase of vacancy concentration) [37] which, adequately modelled, would shift the E_F towards a local minimum. Even if EDXS elementary composition is compatible with a slightly off-stoichiometry, there is no strong indication of this coming from XRD data, so, the stoichiometric model was considered here. Further experimental investigations will be carried out aiming physical properties studies of this compound.

Lu$_5$Pd$_4$Ge$_8$ Lu$_3$Pd$_4$Ge$_4$

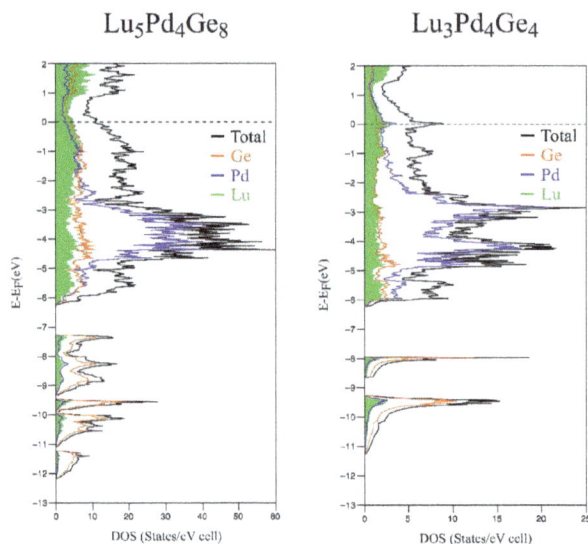

Figure 5. Total and projected DOS for the two studied compounds.

For both compounds, the valence orbital mixing of the three components over the whole energy range is noteworthy. Below E_F, both DOSs showed a gap of around -7 eV separating the two regions, with the lowest being mostly dominated by the *4s* Ge states. The Pd-*d* states are mainly distributed in the range between -5 and -2.5 eV. Their width and energy overlapped with *4p* Ge and Lu states, supporting the bonding relevance of Pd–Ge and Pd–Lu interactions. The fact that the majority of Pd *4d* states are located well below the E_F indicates the electron acceptor character of this species. A significant contribution of *5d* Lu states just below the E_F is a common feature of cations in polar intermetallics, characterized by an incomplete charge transfer (confirmed here also by Bader charge values).

Although the *Zintl–Klemm* (8–N) rule cannot be applied for the title compounds, it was decided to trace interaction similarities comparing the electronic structures of ideal Zintl anions Ge$_2^{6-}$ and *cis*-Ge$_4^{10-}$ coming from the extended Hückel calculation with those obtained by means of TB-LMTO-ASA, in terms of COHP curves. Molecular orbital diagrams (MO) for Ge$_2^{6-}$ (point group $D_{\infty h}$) and Ge$_4^{10-}$ (the point symmetry of this anion was forced to C_{2v} fixing for all distances to 2.56 Å and obtuse internal angles to 111°) are presented in the Supplementary Materials (Figure S2) with the accordingly labeled orbitals.

In Figure 6a, the molecular orbital overlap population (MOOP) for Ge$_2^{6-}$ is shown, together with COHP curves for Ge–Ge interactions (in dumbbells) existing in Lu$_3$Pd$_4$Ge$_4$ and Lu$_5$Pd$_4$Ge$_8$.

These partitioning methods could not be directly compared, since MOOP partitions the electron number, instead, COHP partitions the band structure energy. Since they both permit to easily distinguish between bonding and antibonding states, it was decided to perform a qualitative comparison targeting to figure out the similarities/differences between the isolated molecular fragments analogous with those found in the studied compounds.

The presence of the gap (at ca. -7eV) may be attributed to the energy separation of the σ_{ss} and σ^*_{ss} of Ge$_2$ dumbbells from the σp, πp and π^* orbitals. For the Lu$_3$Pd$_4$Ge$_4$ there are some occupied π^* states close to E_F, whereas in Lu$_5$Pd$_4$Ge$_8$, the cited interactions are almost optimized at E_F. From these observations it derives that Ge dumbbells are not completely polarized; for Lu$_5$Pd$_4$Ge$_8$ the dispersion of σ and σ^* states is more pronounced. One of the possible explanation of this is the existence of additional covalent interactions between germanium dumbbells and neighboring atoms.

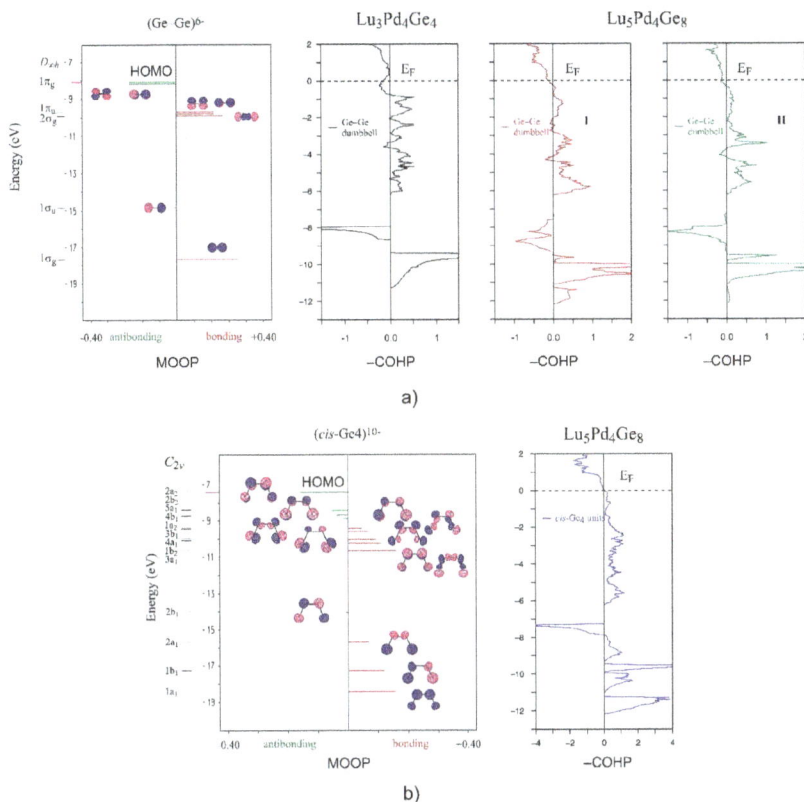

Figure 6. Extended Hückel calculated Molecular Orbital Overlap Population (MOOP) plot for the Ge_2^{6-} (**a**) and *cis*-Ge_4^{10-} (**b**) anions together with the corresponding Crystal Orbital Hamilton Population (COHP) for $Lu_3Pd_4Ge_4$ and $Lu_5Pd_4Ge_8$ (**I** and **II** corresponds to two distinct dumbbells). The degeneracy of the π levels for Ge_2^{6-} is removed for the sake of clarity. The HOMO energy is set in correspondence to E_F.

From the structural data it is known that in $Lu_3Pd_4Ge_4$, Ge atoms are distanced at 2.59 Å as in diverse metal-like salts studied before [38–40]. Instead, in $Lu_5Pd_4Ge_8$ this distance is shortened to 2.49 Å. Usually, the trend of Ge–Ge dumbbell distances is related with electrostatic repulsion between atoms. This statement is coherent with integrated COHP values (–*i*COHP, see Tables 4 and 5) that reflect the same trend, being of −1.82 eV/cell for $Lu_3Pd_4Ge_4$ and of −2.39 and −2.48 eV/cell for $Lu_5Pd_4Ge_8$.

Within the *cis*-Ge_4^{10-} anion the number of covalent interactions is higher, as a result the energy dispersion of its molecular states increases. For example, in the range $-18 \div -14$eV there are four MOs instead of two MOs for dumbbells. A very similar trend/type of interactions derives from COHP curves for $Lu_5Pd_4Ge_8$. As for the dumbbells, the interactions for the *cis* fragment are optimized at the E_F confirming its partial polarization.

Based on –*i*COHP values listed in Tables 4 and 5 it derives that Pd–Ge interactions are very relevant, so one may assume the covalent type of bonding between them. The –COHP plots in Figure 7 confirm that they are mainly of bonding type over a large range below E_F with a weak unfavorable antibonding interaction in the vicinity of E_F, probably due to electrostatic repulsion between Ge orbitals and filled *d* states of Pd.

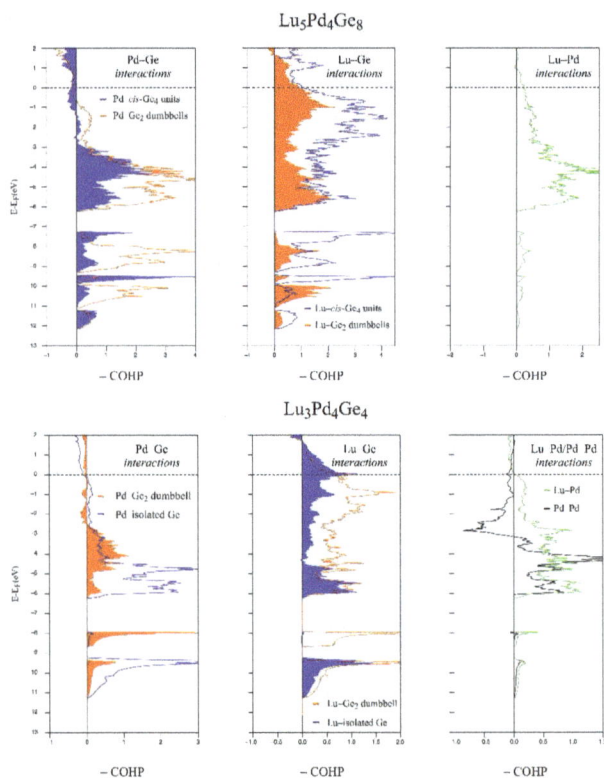

Figure 7. Crystal Orbital Hamilton Populations (–COHP) for selected interactions for the two studied compounds.

Inside $Lu_3Pd_4Ge_4$ the presence of a Pd–Pd short interaction can be highlighted. The –COHP plots for this are similar to those reported for Ca_2Pd_3Ge [41] showing a sharp antibonding character around -3 eV commonly attributed to enhanced repulsion between filled *d* states of Pd. Nevertheless, they are of bonding type in average as deducible from the –iCOHP values for this interaction (0.97 eV/cell), comparable to those reported in [41].

The remaining Lu–Pd and Ge–Lu interactions are weaker being however very similar for both germanides. All of them are of bonding type, Lu–Pd interactions are practically optimized at Fermi level. Numerous interactions between Lu and Pd (Lu and Ge) suggest that some covalent-like interaction may exist due to mixing between *d* states of Lu and Pd (or *d* states of Lu with *p* of Ge; similarly, as it was reported for Ca_5Ge_3 [36] and CaSi [42]). More detailed studies are needed in order to interpret these interactions.

The existence of the complex Pd–Ge polyanion (illustrated in Figure 2) and the electronegativity difference between Pd and Ge explains the trend of Ge species charges (listed in Table 6). The Ge dumbbell in $Lu_3Pd_4Ge_4$ has four neighboring Pd atoms, instead those in $Lu_5Pd_4Ge_8$ install six Pd–Ge polar interactions. As a result, the latter Ge species has lower negative charges. The same is true for (0b)Ge atom with six palladium atoms around in $Lu_3Pd_4Ge_4$: its charge approaches to zero. Within crystal structure, the number of Pd–Ge contacts is the same for terminal and central atoms of *cis*–Ge_4 units; thus, their charges trend is similar as for ideal *cis*-Ge_4^{10-} anion, terminal atoms being more negative.

4. Conclusions

The two new Lu$_5$Pd$_4$Ge$_8$ and Lu$_3$Pd$_4$Ge$_4$ polar intermetallics were synthesized and characterized in this work. They were found to crystallize in the *mP*34–Tm$_5$Pd$_4$Ge$_8$ and *oI*22–Gd$_3$Cu$_4$Ge$_4$ structures respectively. A detailed description of crystal structure solution in the case of the non-merohedral twinned crystal of Lu$_5$Pd$_4$Ge$_8$ was proposed, highlighting the difficulties/problems encountered here along with practical suggestions to manage them.

Joined crystal chemical analysis and combined DFT studies suggest the presence of [Pd$_4$Ge$_8$]$^{7.4-}$ and [Pd$_4$Ge$_4$]$^{4.6-}$ polyanions. The interactions of Lu with these frameworks cannot be viewed as purely ionic as derives from its states distribution, COHP analysis and Bader charges. The Lu–Pd and Lu–Ge bonding interactions are one of the most interesting aspects arisen from our study and their nature deserves further investigations.

Supplementary Materials: The following are available online at http://www.mdpi.com/2073-4352/8/5/205/s1, Figure S1: Schematic representation of the structural relationships between "cis" and "trans" Ge$_4$ fragments in Lu$_5$Pd$_4$Ge$_8$ models; Figure S2: Molecular orbitals diagram for Ge$_2$$^{6-}$ (a) and *cis*-Ge$_4$$^{10-}$ (b) as generated by CACAO; Figure S3: Total DOS for Lu$_5$Pd$_4$Ge$_8$ and Lu$_3$Pd$_4$Ge$_4$ together with the orbital projected DOS for each species; Table S1: Atomic parameters for "trans"-Lu$_5$Pd$_4$Ge$_8$ model. Video S1: Lu$_5$Pd$_4$Ge$_8$_trans-cis_optimization, Lu$_5$Pd$_4$Ge$_8$ CIF file, Lu$_3$Pd$_4$Ge$_4$ CIF file.

Author Contributions: Riccardo Freccero, Pavlo Solokha and Serena De Negri conceived and designed the experiments; Riccardo Freccero performed the syntheses; Pavlo Solokha and Davide Maria Proserpio performed the XRD single crystal experiments; Serena De Negri performed SEM analyses; Riccardo Freccero, Pavlo Solokha and Davide Maria Proserpio conducted different calculations; Riccardo Freccero, Pavlo Solokha, Serena De Negri and Adriana Saccone analyzed the data and wrote the paper.

Acknowledgments: The authors thank Roman Eremin from SCTMS (Samara State University, Russia) for his contribution in the Bader charge analysis.

Conflicts of Interest: The authors declare no conflict of interest.

References

1. Villars, P.; Cenzual, K. *Pearson's Crystal Data*; ASM International: Metals Park, OH, USA, 2018.
2. Freccero, R.; Solokha, P.; Proserpio, D.M.; Saccone, A.; De Negri, S. A new glance on R$_2$MGe$_6$ (R = rare earth metal, M = another metal) compounds. An experimental and theoretical study of R$_2$PdGe$_6$ germanides. *Dalton Trans.* **2017**, *46*, 14021–14033. [CrossRef] [PubMed]
3. Niepman, D.; Prots, Y.M.; Pöttgen, R.; Jeitschko, W. The order of the palladium and germanium atoms in the germanides *Ln*PdGe (*Ln* = La–Nd, Sm, Gd, Tb) and the new compound Yb$_3$Pd$_4$Ge$_4$. *J. Solid State Chem.* **2000**, *154*, 329–337. [CrossRef]
4. Heying, B.; Rodewald, U.C.; Pöttgen, R. The germanides Er$_5$Pd$_4$Ge$_8$ and Tm$_5$Pd$_4$Ge$_8$—3D [Pd$_4$Ge$_8$] polyanions with Ge$_2$ dumb-bells and Ge$_4$ chains in *cis*-conformation. *Z. Kristallogr.* **2017**, *232*, 435–440. [CrossRef]
5. Feyerherm, R.; Becker, B.; Collins, M.F.; Mydosh, J.; Nieuwenhuys, G.J.; Ramakrishnan, S. The magnetic structure of CePd$_2$Ge$_2$ and Ce$_2$Pd$_3$Ge$_5$. *Phys. B* **1998**, *241–243*, 643–645. [CrossRef]
6. Anand, V.K.; Thamizhavel, A.; Ramakrishnan, S.; Hossain, Z. Complex magnetic order in Pr$_2$Pd$_3$Ge$_5$: A single crystal study. *J. Phys.* **2012**, *24*. [CrossRef] [PubMed]
7. Solokha, P.; Freccero, R.; De Negri, S.; Proserpio, D.M.; Saccone, A. The R$_2$Pd$_3$Ge$_5$ (R = La–Nd, Sm) germanides: Synthesis, crystal structure and symmetry reduction. *Struct. Chem.* **2016**, *27*, 1693–1701. [CrossRef]
8. Rodewald, U.C.; Heying, B.; Hoffmann, R.-D.; Niepmann, D. Polymorphism in the germanides REPdGe with the heavy rare earth elements. *Z. Naturforsch.* **2009**, *64b*, 595–602. [CrossRef]
9. Chabot, B.; Parthé, E. Dy$_2$Co$_3$Si$_5$, Lu$_2$Co$_3$Si$_5$, Y$_2$Co$_3$Si$_5$ and Sc$_2$Co$_3$Si$_5$ with a monoclinic structural deformation variant of the orthorhombic U$_2$Co$_3$Si$_5$ structure type. *J. Less-Common Met.* **1985**, *106*, 53–59. [CrossRef]
10. Bruker. APEX2, SAINT-Plus, XPREP, SADABS, CELL_NOW and TWINABS. Bruker AXS Inc.: Madison, WI, USA, 2014.
11. Kraus, W.; Nolze, G. POWDER CELL-A program for the representation and manipulation of crystal structure and calculation of the resulting X-ray powder pattern. *J. Appl. Crystallogr.* **1996**, *29*. [CrossRef]

12. Bader, R.F. *Atoms in Molecules: A Quantum Theory*; Clarendon Press and Oxford University Press: New York, NY, USA, 1994.

13. Kresse, G.; Furthmüller, J. Efficient iterative schemes for ab initio total-energy calculations using a plane-wave basis set. *J. Phys. Rev. B* **1996**, *54*, 11169–11186. [CrossRef]

14. Krier, G.; Jepsen, O.; Burkhardt, A.; Andersen, O.K. *The TB-LMTO-ASA Program*; Version 4.7; Max-Planck-Institut Für Festkörperforschung: Stuttgart, Germany, 2000.

15. Barth, U.; Hedin, L.A. Local exchange-correlation potential for the spin polarized case: I. *J. Phys. Chem.* **1972**, *C5*, 1629–1642. [CrossRef]

16. Dronskowski, R.; Blöchl, P.E. Crystal orbital Hamilton populations (COHP): Energy-resolved visualization of chemical bonding in solids based on density-functional calculations. *J. Phys. Chem.* **1993**, *97*, 8617–8624. [CrossRef]

17. Eck, B. Wxdragon, Aachen, Germany, 1994–2018. Available online: http://wxdragon.de/ (accessed on 2 May 2018).

18. Mealli, C.; Proserpio, D.M. MO Theory made visible. *J. Chem. Educ.* **1990**, *67*, 399–403. [CrossRef]

19. Mealli, C.; Ienco, A.; Proserpio, D.M. *Book of Abstracts of the XXXIII. ICCC*; ICCC: Florence, Italy, 1998.

20. Lowe, J.; Peterson, K. *Quantum Chemistry*, 3rd ed.; Academic Press: Cambridge, MA, USA, 2005.

21. Hoffmann, R. *Solids and Surfaces: A Chemist's View of Bonding in Extended Structures*; VCH: New York, NY, USA, 1988.

22. Solokha, P.; De Negri, S.; Saccone, A.; Proserpio, D.M. On a non-merohedrally twinned Tb_3Ge_5 crystal. In Proceedings of the XII International Conference on Crystal Chemistry of Intermetallic Compounds, Lviv, Ukraine, 22–26 September 2012.

23. Budnyk, S.L.; Weitzer, F.; Kubata, C.; Prots, Y.; Akselrud, L.G.; Schnelle, W.; Hiebl, K.; Nesper, R.; Wagner, F.R.; Grin, Yu. Barrelane-like germanium clusters in Eu_3Ge_5: Crystal structure, chemical bonding and physical properties. *J. Solid State Chem.* **2006**, *179*, 2329–2338. [CrossRef]

24. Shcherban, O.; Savysyuk, I.; Semuso, N.; Gladyshevskii, R.; Cenzual, K. Crystal structure of the compound Pr_4Ge_7. *Chem. Met. Alloys* **2009**, *2*, 115–122.

25. Müller, P. *Crystal Structure Refinement: A Crystallographer's Guide to SHELXL*; Oxford University Press: Oxford, UK, 2006.

26. Clegg, W. *Crystal Structure Analysis: Principle and Practice*, 2nd ed.; Oxford University Press: Oxford, UK, 2009.

27. Petricek, V.; Dusek, M.; Palatinus, L. Crystallographic computing system JANA2006: General features. *Z. Kristallogr.* **2014**, *229*, 345–352.

28. Cooper, R.I.; Gould, R.O.; Parsons, S.; Watkin, D.J. The derivation of non-merohedral twin laws during refinement by analysis of poorly fitting intensity data and the refinement of non-merohedrally twinned crystal structures in the program crystals. *J. Appl. Crystallogr.* **2002**, *35*, 168–174. [CrossRef]

29. Farrugia, L.J. WinGX and ORTEP for windows: An update. *J. Appl. Cryst.* **2012**, *45*, 849–854. [CrossRef]

30. Gelato, L.; Parthé, E. *Structure tidy*—A computer program to standardize crystal structure data. *J. Appl. Crystallogr.* **1987**, *20*, 139–143. [CrossRef]

31. Parthé, E.; Gelato, L.; Chabot, B.; Penzo, M.; Cenzual, K.; Gladyshevskii, R. *TYPIX Standardized Data and Crystal Chemical Characterization of Inorganic Structure Types*; Springer-Verlag: Heidelberg, Germany, 1993.

32. Prots, Y.; Demchyna, R.; Burkhardt, U.; Schwarz, U. Crystal structure and twinning of HfPdGe. *Z. Kristallogr.* **2007**, *222*, 513–520. [CrossRef]

33. Wondratschek, H.; Müller, U. *International Tables for Crystallography, Symmetry Relations between Space Groups*; Vol. A1; Kluwer Academic Publishers: Dordrecht, The Netherlands, 2004.

34. Müller, U. *Symmetry Relationships between Crystal Structures. Applications of Crystallographic Group Theory in Crystal Chemistry*; Oxford University Press: Oxford, UK, 2013.

35. Pöttgen, R. Coloring, distortions, and puckering in selected intermetallic structures from the perspective of group—Subgroup relations. *Z. Anorg. Allg. Chem.* **2014**, *640*, 869–891. [CrossRef]

36. Landrum, G.A.; Dronskowski, R. The orbital origins of magnetism: from atoms to molecules to ferromagnetic alloys. *Angew. Chem. Int. Ed.* **2000**, *39*, 1560–1585. [CrossRef]

37. Lin, Q.; Corbett, J.D. Centric and non-centric $Ca_3Au_{\sim 7.5}Ge_{\sim 3.5}$: Electron-poor derivatives of La_3Al_{11}. syntheses, structures, and bonding analyses. *Inorg. Chem.* **2009**, *48*, 5403–5411. [CrossRef] [PubMed]

38. Mudring, A.-V.; Corbett, J.D. Unusual electronic and bonding properties of the zintl phase Ca_5Ge_3 and related compounds. A theoretical analysis. *J. Am. Chem. Soc.* **2004**, *126*, 5277–5281. [CrossRef] [PubMed]

39. Siggelkow, L.; Hlukhyy, V.; Fässler, T.F. Sr_7Ge_6, Ba_7Ge_6 and Ba_3Sn_2—Three new binary compounds containing dumbbells and four-membered chains of tetrel atoms with considerable Ge-Ge π-bonding character. *J. Solid State Chem.* **2012**, *191*, 76–89. [CrossRef]

40. Eisenmann, B.; Schäfer, H. Zur Strukturchemie der Verbindungsreihe $BaMg_2X_2$ (X = Si, Ge, Sn, Pb). *Z. Anorg. Allg. Chem.* **1974**, *403*, 163–172. [CrossRef]

41. Doverbratt, I.; Ponou, S.; Lidin, S. Ca_2Pd_3Ge, a new fully ordered ternary Laves phase structure. *J. Solid State Chem.* **2013**, *197*, 312–316. [CrossRef]

42. Kurylyshyn, I.M.; Fässler, T.F.; Fischer, A.; Hauf, C.; Eickerling, G.; Presnitz, M.; Scherer, W. Probing the Zintl-Klemm concept: A combined experimental and theoretical charge density study of the Zintl phase CaSi. *Angew. Chem. Int. Ed.* **2014**, *53*, 3029–3032. [CrossRef] [PubMed]

crystals

MDPI

Article

Mixed Sr and Ba Tri-Stannides/Plumbides $A^{II}(Sn_{1-x}Pb_x)_3$

Michael Langenmaier [†,‡], **Michael Jehle** [†,‡] and **Caroline Röhr** *

Institut für Anorganische und Analytische Chemie, Albert-Ludwigs-Universität Freiburg, 79104 Freiburg, Germany; michil@limonite.chemie.uni-freiburg.de (M.L.); michij@almandine.chemie.uni-freiburg.de (M.J.)
* Correspondence: caroline@ruby.chemie.uni-freiburg.de; Tel.: +49-0761-203-6143
† These authors contributed equally to this work.
‡ Institut für Anorganische und Analytische Chemie, Albert-Ludwigs-Universität Freiburg, Albertstr. 21, D-79104 Freiburg, Germany.

Received: 2 April 2018 ; Accepted: 26 April 2018; Published: 4 May 2018

Abstract: The continuous substitution of tin by lead (M^{IV}) allows for the exploration geometric criteria for the stability of the different stacking variants of alkaline-earth tri-tetrelides $A^{II}M_3^{IV}$. A series of ternary Sr and Ba mixed tri-stannides/plumbides $A^{II}(Sn_{1-x}Pb_x)_3$ (A^{II} = Sr, Ba) was synthesized from stoichiometric mixtures of the elements. Their structures were determined by means of single crystal X-ray data. All structures exhibit close packed ordered AM_3 layers containing M kagomé nets. Depending on the stacking sequence, the resulting M polyanion resembles the oxygen substructure of the hexagonal (face-sharing octahedra, h stacking, Ni_3Sn-type, border compound $BaSn_3$) or the cubic (corner-sharing octahedra, c stacking, Cu_3Au-type, border compound $SrPb_3$) perovskite. In the binary compound $BaSn_3$ (Ni_3Sn-type) up to 28% of Sn can be substituted against Pb ($hP8$, $P6_3/mmc$, $x = 0.28(4)$: $a = 726.12(6)$, $c = 556.51(6)$ pm, R1 = 0.0264). A further increased lead content of 47 to 66% causes the formation of the $BaSn_{2.57}Bi_{0.43}$-type structure with a $(hhhc)_2$ stacking [$hP32$, $P6_3/mmc$, $x = 0.47(3)$: $a = 726.80(3)$, $c = 2235.78(14)$ pm, R1 = 0.0437]. The stability range of the $BaPb_3$-type sequence $(hhc)_3$ starts at a lead proportion of 78% ($hR36$, $R\bar{3}m$, $a = 728.77(3)$, $c = 2540.59(15)$ pm, R1 = 0.0660) and reaches up to the pure plumbide $BaPb_3$. A second new polymorph of $BaPb_3$ forms the Mg_3In-type structure with a further increased amount of cubic sequences [$(hhcc)_3$; $hR48$, $a = 728.7(2)$, $c = 3420.3(10)$ pm, R1 = 0.0669] and is thus isotypic with the border phase $SrSn_3$ of the respective strontium series. For the latter, a Pb content of 32% causes a small existence region of the $PuAl_3$-type structure [$hP24$, $P6_3/mmc$, $a = 696.97(6)$, $c = 1675.5(2)$ pm, R1 = 0.1182] with a $(hcc)_2$ stacking. The series is terminated by the pure c stacking of $SrPb_3$, the stability range of this structure type starts at 75% Pb ($cP4$, $Pm\bar{3}m$; $a = 495.46(9)$ pm, R1 = 0.0498). The stacking of the close packed layers is evidently determined by the ratio of the atomic radii of the contributing elements. The Sn/Pb distribution inside the polyanion ('coloring') is likewise determined by size criteria. The electronic stability ranges, which are discussed on the basis of the results of FP-LAPW band structure calculations are compared with the *Zintl* concept and *Wade*'s/*mno* electron counting rules. Still, due to the presence of only partially occupied steep M-p bands the compounds are metals exhibiting pseudo band gaps close to the Fermi level. Thus, this structure family represents an instructive case for the transition from polar ionic/covalent towards (inter)metallic chemistry.

Keywords: stannides; plumbides; alkaline-earth

---

1. Introduction

In view of the complex bonding situation between ionic, covalent and metallic, the synthetic, crystallographic and bond theoretical studies on geometric and electronic parameters determining the structure chemistry (and therewith also the properties) of polar intermetallics of the alkali and alkaline-earth compounds of the *p*-block elements are still a fascinating field of research. Compared to the vast number of alkali/alkaline-earth stannides and plumbides [1], which are simple electron-precise *Zintl* phases, the tri-stannides and -plumbides ASn_3 and APb_3 of the alkaline-earth elements (A = Ca, Sr, Ba) are ordered derivatives of simple dense packings of the two elements [2–9]. Similar to the variation of the alkaline-earth elements (A) in the pure tri-stannides [10] and tri-plumbides [11], a continuous substitution of tin by lead allows for the exploration of geometric and electronic (Sn/Pb distribution) criteria for the stability of different stacking variants of the hexagonal close packed layers [AM_3] in these tri-tetrelides $A^{II}M^{IV}{}_3$.

The tri-plumbides APb_3 of calcium and strontium, which both form the cubic Cu_3Au-type structure ($\overline{|:ABC:|}$ or *c* stacking after *Jagodzinski*), have been structurally characterized using X-ray powder data in the past and some physical properties as well as the electronic band structure were reported in the 1960s and 1970s by Havinga et al. [6,12,13]. More recent works on their electronic and magnetic properties were published by Baranovskiy et al. [14,15]. The structure of the binary barium plumbide $BaPb_3$ [$(hhc)_3$ stacking] was derived and refined from 78 reflections by *Sands, Wood* and *Ramsey* as early as 1964 [9]. Only two years later, *van Vucht* published an elaborate work [7], in which he, based on powder diffraction data, already suggested the existence of further stacking variants for the mixed alkaline-earth plumbides $A_xBa_{1-x}Pb_3$ (A = Ca, Sr). These two series have been fully explored by our group 10 years ago [11]: Both the Ca and the Sr series start with the pure *c* stacking of $(Ca/Sr)Pb_3$. At an approximate 1:1 ratio of strontium and barium (e.g., 35 to 53% Sr) the $PuAl_3$-type [$(hcc)_2$ stacking] has a distinct homogeneity range. Reversely, the $(hhc)_3$ stacking of the binary compound $BaPb_3$ changes at an already very small partial substitution of barium against calcium ($Ca_{0.03}Ba_{0.97}Pb_3$) or strontium ($Sr_{0.11}Ba_{0.89}Pb_3$) towards the $(hhcc)_3$ sequence of the Mg_3In-type.

For the tri-stannides ASn_3, the continuous substitution of the different alkaline-earth cations against each other has been likewise investigated in a systematic study [10]: Starting from $CaSn_3$ (pure *c* stacking [2]) up to 46% of Ca can be substituted against Sr without a structural change. The binary phase $SrSn_3$ forms the Mg_3In-type structure with a $(hhcc)_3$ stacking sequence [3]. A small partial substitution of Sr against Ba (9 to 19%) causes the packing to switch to the $(hhc)_3$ sequence of the $BaPb_3$-type. At a Ba proportion of 26% a further structure change to the $BaSn_{2.57}Bi_{0.43}$-type structure [$(hhhc)_2$ stacking] takes place. The tin series terminates with the pure *h* stacking of $BaSn_3$; the stability range of this Ni_3Sn-type structure starts at the composition $Sr_{0.22}Ba_{0.78}Sn_3$.

Beyond these pure geometric/size effects, *van Vucht* and *Havinga* [8,16] also demonstrated the electronic influence on the *h/c* stacking of the layers in several Tl-substituted plumbides $A(Tl/Pb)_3$. An indium content in $Ba(In/Pb)_3$ of 15% also changes the stacking from the simple cubic to the mixed $(hcc)_2$ sequence of the $PuAl_3$-type [17]. For the electron-richer system $BaSn_3$–$BaBi_3$ a similar influence of the valence electron (v.e.) number on the stacking was shown more recently by Fässler et al. [18].

In the present work, we report on the results of a systematic experimental, crystallographic and bond-theoretical DFT investigation of mixed Sn/Pb tri-tetrelides of Sr and Ba. Here, the substitution of tin by lead does not only change the geometric relations, but in addition the occupation of the crystallographically different *M* positions of the more complex stacked sequences with tin and/or lead, i.e., the 'coloring' of the polyanion [19] can be studied.

2. Experimental

2.1. Synthesis and Phase Widths

The synthesis of the mixed Sn/Pb tri-tetrelides was generally performed starting from the elements strontium or barium, tin and lead as obtained from commercial sources (Sr, Ba: Metallhandelsgesellschaft Maassen, Bonn, 99%; Sn: shots, 99.9%, Riedel de Häen; Pb: powder, 99.9%, Merck KGaA). The elements were filled into tantalum crucibles in a glovebox under an argon atmosphere and the sealed containers were heated up with a rate of 200 °C/h to maximum temperatures (T_{max}) of 700 to 800 °C. For all samples, the crystallization was performed by a slow cooling rate ($\dot{T}^{\downarrow}$) of 2 to 20 °C/h. The maximum temperatures applied and the detailed cooling rates can be found in Table 1, together with the weighed sample compositions, which were mainly restricted to the stoichiometric 1:3 [for (Sr/Ba):(Sn/Pb)] element ratio. All sample reguli were hard and brittle with a dark-metallic luster and are sensitive against moisture. Representative parts of the reguli were ground and sealed in capillaries with a diameter of 0.3 mm. X-ray powder patterns were collected on transmission powder diffraction systems (STADI-P or Dectris Mythen 1K detector, Stoe & Cie, Darmstadt, MoK_{α} radiation, graphite monochromator). For the phase analysis, the measured powder patterns were compared to the calculated (program LAZY-PULVERIX [20]) reflections of the title compounds and other known phases in the respective ternary systems.

Table 1. Details of the synthesis of the title compounds (cf. Tables 2 and 3 for the single crystal numbers; uniform heating rate: $\dot{T}^{\uparrow}$ = 200 °C/h) .

Sample	Structure	Single	Pb	Weighed Elements						Temperature Program					
Composition	Type	Crystal	Content	Sr/Ba		Sn		Pb		T_{max}	$\dot{T}^{\downarrow}$	T	$\dot{T}^{\downarrow}$	T	$\dot{T}^{\downarrow}$
		No.	[%]	[mg]	[mmol]	[mg]	[mmol]	[mg]	[mmol]	(in [°C] and [°C/h])					
BaSn$_2$Pb	Ni$_3$Sn	1	28	253.7	1.85	407.0	3.43	356.8	1.72	750	30	500	200		
BaSn$_{1.5}$Pb$_{1.5}$	BaSn$_{2.6}$Bi$_{0.4}$	2	47.4	220.7	1.61	284.7	2.40	496.7	2.40	750	200	650	2	500	200
BaSnPb$_2$	BaSn$_{2.6}$Bi$_{0.4}$	3	66.3	205.2	1.49	176.0	1.48	619.4	2.99	750	20	500	200		
BaSn$_{0.75}$Pb$_{2.25}$	BaPb$_3$	4	78.3	199.6	1.45	129.1	1.09	675.9	3.26	750	200	650	2	500	200
BaSn$_{0.5}$Pb$_{2.5}$	BaPb$_3$	4p	85	192.2	1.40	88.1	0.74	724.3	3.50	750	30	500	200		
BaGePb$_2$	Mg$_3$In	5	100	220.6	1.61	116.9	1.61	663.6	3.20	750	30	500	200		
SrSn$_{2.5}$Pb$_{0.5}$	Mg$_3$In	5p	14	179.7	2.05	608.8	5.13	212.3	1.02	800	200	750	2	625	200
SrSn$_2$Pb	PuAl$_3$	6	32	164.8	1.88	445.5	3.75	389.6	1.88	800	200	750	2	625	200
SrSn$_{1.5}$Pb$_{1.5}$	PuAl$_3$	6p	34	152.0	1.73	307.9	2.59	539.0	2.60	700	20	500	200		
SrSn$_{0.75}$Pb$_{2.25}$	Cu$_3$Au	7	75	137.1	1.56	138.7	1.17	726.0	3.50	700	20	500	200		

Starting from the pure tin compound, the first ternary sample of the <u>Ba series</u>, with an overall composition of BaSn$_2$Pb, yielded the border phase BaSn$_{2.16}$Pb$_{0.84}$ (**1**) of the Ni$_3$Sn-type (magenta bar in Figure 1) in pure phase. The powder pattern of the samples with a Pb content of 50% and 66.7% could be fully indexed with the data of the BaSn$_{2.6}$Bi$_{0.4}$-type. Accordingly, the two border phases of the existence range of this structure type (dark blue bar in Figure 1), BaSn$_{1.58}$Pb$_{1.42}$ (**2**) and BaSn$_{1.01}$Pb$_{1.99}$ (**3**), were obtained from these two samples. The following two samples with a further increased lead content, BaSn$_{0.75}$Pb$_{2.25}$ and BaSn$_{0.5}$Pb$_{2.5}$, yielded ternary Sn-containing variants of the known BaPb$_3$-type structure (cyan). The refined chemical composition of the border phase obtained from the first sample, BaSn$_{0.65}$Pb$_{2.35}$ (**4**), is very close to the weighed sample stoichiometry, the composition of the somewhat Pb-richer phase was refined from powder data. A second form of BaPb$_3$, which is isostructural to SrSn$_3$ (green bar in Figure 1), was obtained for the first time from the Ge-containing sample (BaGePb$_2$), which contains (optically visible) elemental germanium as a second product. In the Sr series, a lead content of 17% (SrSn$_{2.5}$Pb$_{0.5}$) still yielded the Mg$_3$In-type structure of the binary tri-stannide SrSn$_3$. The composition of this phase (SrSn$_{2.58}$Pb$_{0.42}$) was refined from powder data (Section 2.3). The two samples with an Sn-content of 66 and 50% contain PuAl$_3$-type phases. For the first sample, the single crystal data of selected crystals are in very good agreement with the weighed element ratio (**6**: SrSn$_{2.05}$Pb$_{0.95}$, Section 2.2). Due to the low crystal quality, the composition of the Pb-richer phase of this type could be only refined from powder data (Section 2.3). A small content of the Pb-richer cubic Cu$_3$Au-type phase, which is formed in X-ray pure form in the final sample SrSn$_{0.75}$Pb$_{2.25}$, could not be excluded due to the overlap of the few reflections of the Cu$_3$Au-type with those of the structurally related PuAl$_3$-type.

2.2. Single Crystal Structure Refinements

For the crystal structure refinements (cf. Tables 2 and 3), irregularly shaped single crystals were selected using a stereo microscope and mounted in glass capillaries (diameter 0.1 mm) under dried paraffine oil. The crystals were centered on diffractometers equipped with an image plate (Stoe IPDS, sealed tube) or a CCD detector (Bruker Apex-II Quazar, microfocus source).

Table 2. Crystallographic data and details of the data collection and structure refinement of the compounds of the barium series $BaSn_{2.16}Pb_{0.84}$ (**1**), $BaSn_{1.58}Pb_{1.42}$ (**2**), $BaSn_{1.01}Pb_{1.99}$ (**3**), $BaSn_{0.65}Pb_{2.35}$ (**4**), and $BaPb_3$ (**5**) and the strontium compounds $SrSn_{2.05}Pb_{0.95}$ (**6**) and $SrSn_{0.75}Pb_{2.25}$ (**7**).

	A^{II}	Ba Compounds					Sr Compounds	
No.		1	2	3	4	5	6	7
Pb content, %		28	47.4	66.3	78.3	100	32	75
Structure type		Ni_3Sn	$BaSn_{2.6}Bi_{0.4}$		$BaPb_3$	Mg_3In	$PuAl_3$	Cu_3Au
Stacking sequence		h_2	$(hhhc)_2$		$(hhc)_3$	$(hhcc)_3$	$(hcc)_2$	c_3
Crystal system			hexagonal		rhombohedral		hexagonal	cubic
Space group			$P6_3/mmc$		$R\bar{3}m$		$P6_3/mmc$	$Pm\bar{3}m$
			no. 194		no. 166		no. 194	no. 221
Pearson symbol		$hP8$	$hP32$		$hR36$		$hR48$	$cP4$
Lattice parameters, pm	a	726.12(6)	726.80(3)	728.12(4)	728.77(3)	728.7(2)	696.97(6)	495.46(9)
	c	556.51(6)	2235.78(14)	2254.81(13)	2540.6(2)	3420.3(12)	1675.5(2)	
Volume of the u.c., 10^6 pm^3		254.11(5)	1022.80(11)	1035.25(13)	1168.55(12)	1572.8(11)	704.87(15)	
Volume/f.u., 10^6 pm^3		127.06	127.85	129.41	129.84	131.07	117.49	121.63(7)
c/a		1.5328	1.5380	1.5484	1.5494	1.5646	1.6027	1.63
Z		2	8	8	9	12	6	1
Density (X-ray), gcm^{-3}		7.38	8.04	8.59	8.97	9.62	7.46	8.78
Diffractometer			Stoe IPDS-II				APEX-II	
				Mo-K_α radiation				
Absorption coefficient $\mu_{Mo-K\alpha}$, mm^{-1}		44.7	61.7	76.6	86.2	103.3	55.7	92.1
range, deg		3.2–29.1	1.8–29.3	1.8–29.3	2.4–29.1	1.8–30.0	3.4–29.9	4.1–32.5
No. of reflections collected		3737	7730	9690	5106	5074	6364	730
No. of independent reflections		155	582	593	431	628	433	69
R_{int}		0.0723	0.0732	0.1072	0.2342	0.1131	0.1023	0.0860
Corrections			Lorentz, Polarisation, Absorption					
			XShape [21]				Sadabs [22]	
Structure refinement				SHELXL-2013 [23]				
No. of free parameter		9	25	25	18	20	18	6
Goodness-of-fit on F^2		1.248	1.372	1.383	1.116	0.969	1.243	1.263
R Values [for refl. with $I \geq 2\sigma(I)$]	$R1$	0.0264	0.0437	0.0423	0.0660	0.0499	0.1182	0.0571
	$wR2$	0.0644	0.0934	0.0566	0.1738	0.1166	0.3585	0.1610
R Values (all data)	$R1$	0.0266	0.0485	0.0506	0.0683	0.0772	0.1270	0.0572
	$wR2$	0.0644	0.0948	0.0578	0.1761	0.1258	0.3642	0.1610
Residual elect. density, $e^- \times 10^{-6}$ pm^{-3}		+1.2/−1.5	+2.2/−1.8	+2.7/−1.8	+5.6/−1.9	+6.4/−3.1	+9.8/−3.7	+7.4/−4.0

The reflections of the tin-rich compounds of the Ba series could be indexed by the expected small hexagonal cell of the Ni_3Sn type (space group $P6_3/mmc$). The border compound of the stability range of this structure type, $BaSn_{2.16}Pb_{0.84}$ (**1**), was obtained in pure phase from a sample of overall composition $BaSn_2Pb$. The parameters of the structure model of the pure tri-stannide [5] were refined using the program SHELXL-2013 [23], and the mixed Sn/Pb position was treated with constrained positional and anisotropic displacement parameters (ADPs). The obtained formula $BaSn_{2.16}Pb_{0.84}$ is only slightly Sn-richer than the sample composition. Crystals yielded from the lead-richer samples $BaSn_{1.58}Pb_{1.42}$ and $BaSn_{1.01}Pb_{1.99}$ show a similar hexagonal basis and the same extinction conditions, but the length of the hexagonal c axis increases to about 2200 pm, which indicates the formation of the $BaSn_{2.6}Bi_{0.4}$-type structure [18]. The two border compounds of the stability range of this structure type contain 47.4 (**2**) and 66.3% (**3**) lead, which is again close to the respective sample compositions. Accordingly, the powder patterns of these two samples are fully indexed with the data of the $BaSn_{2.6}Bi_{0.4}$-type.

Table 3. Atomic coordinates and equivalent isotropic displacement parameters (pm^2) for the single crystal structures of the title compounds.

Compound	No.	Atoms	Wyckoff Position	Point Group Symmetry	Stacking Sequence	Pb Prop. /%	x	y	z	$U_{equiv.}$
BaSn$_{2.16}$Pb$_{0.84}$	1	Ba	2d	$\bar{6}m2$			1/3	2/3	3/4	263(6)
		M	6h	mm2	h	28.0(14)	0.14395(7)	2x	1/4	244(4)
BaSn$_{1.58}$Pb$_{1.42}$	2	Ba(1)	2a	3m.			0	0	0	250(6)
		Ba(2)	2b	$\bar{6}m2$			0	0	1/4	201(5)
		Ba(3)	4f	3m.			1/3	2/3	0.13534(8)	176(4)
		M(1)	6h	mm2	h	9.5(13)	0.52781(12)	2x	1/4	202(6)
		M(2)	12k	.m.	h	40.1(12)	0.18965(8)	2x	0.62443(4)	227(4)
		Pb(3)	6g	$.\frac{2}{m}.$	c		1/2	0	0	266(3)
BaSn$_{1.01}$Pb$_{1.99}$	3	Ba(1)	2a	3m.			0	0	0	268(5)
		Ba(2)	2b	$\bar{6}m2$			0	0	1/4	211(5)
		Ba(3)	4f	3m.			1/3	2/3	0.13465(6)	189(4)
		M(1)	6h	mm2	h	23.7(1)	0.52645(9)	2x	1/4	210(4)
		M(2)	12k	.m.	h	70.6(9)	0.18753(5)	2x	0.62418(3)	232(2)
		Pb(3)	6g	$.\frac{2}{m}.$	c		1/2	0	0	262(2)
BaSn$_{0.65}$Pb$_{2.35}$	4	Ba(1)	3a	3m			0	0	0	374(9)
		Ba(2)	6c	3m			0	0	0.21602(10)	324(7)
		M(1)	18h	.m	h	67(2)	0.47707(12)	$-x$	0.22315(4)	364(6)
		Pb(2)	9e	$.\frac{2}{m}$	c		1/2	0	0	388(6)
BaPb$_3$	5	Ba(1)	6c	3m			0	0	0.13034(7)	220(5)
		Ba(2)	6c	3m			0	0	0.28912(7)	234(5)
		Pb(1)	18h	.m	h		0.47742(8)	$-x$	0.12358(3)	252(3)
		Pb(2)	18h	.m	c		0.50625(8)	$-x$	0.29199(3)	297(3)
SrSn$_{2.05}$Pb$_{0.95}$	6	Sr(1)	2b	$\bar{6}m2$			0	0	1/4	180(20)
		Sr(2)	4f	3m.			1/3	2/3	0.0930(5)	185(18)
		Sn(1)	6h	mm2	h		0.5218(4)	0.0436(8)	1/4	182(15)
		M(2)	12k	.m.	c	47(4)	0.16833(19)	0.3367(4)	0.58108(13)	221(11)
SrSn$_{0.75}$Pb$_{2.25}$	7	Sr	1a	m$\bar{3}$m			0	0	0	500(20)
		M	3c	$\frac{4}{m}$m.m	c	75	0	1/2	1/2	159(9)

The diffraction images of the crystals obtained from the Pb-rich sample BaSn$_{0.5}$Pb$_{2.5}$ could be indexed by a rhombohedral lattice with the familiar unit cell basis of 730 pm, but now with a *c* axis of $\approx$2500 pm, which indicates the formation of the BaPb$_3$-type structure. During the single crystal refinement of this structure model, the 18*h* position M(1) turned out to be statistically occupied by tin and the final refinement yielded a Sn occupation of 67(2)%, i.e., an overall composition of BaSn$_{0.65}$Pb$_{2.35}$ (**4**), which is again close to the weighed element ratios. For the likewise rhombohedral unit cell of the second form of BaPb$_3$-II (**5**), the large *c* axis indicated the 12-layer structure of the Mg$_3$In-type, which was also found for ternary derivatives of BaPb$_3$ containing small amounts of Ca or Sr [11]. The structure refinement converged to *R*1 values of 0.05 and the refinement as well as the unit cell volume (cf. the comparison in Section 3.1.4) gave no indication for an incorporation of germanium into the structure. All crystals of the Sr-series were of much lower quality than those of the Ba compounds and the diffraction images of crystals of different sizes all exhibit very broad reflections. Only the structure of the compound **6** (SrSn$_{2.05}$Pb$_{0.95}$), which forms the hexagonal PuAl$_3$-type, could be refined from single crystal data. However, a quite large value of *R*1 and the residual electron density of about +10 e$^-$10^{-6} pm^{-3}, which lies in the vicinity of the M(2) position, result from the unsatisfactory quality of the diffraction data. In the case of the single crystals yielded from the sample SrSn$_{0.75}$Pb$_{2.25}$ (**7**) the situation is even worse, and the Sn:Pb ratio of the sole M position in the cubic Cu$_3$Au-type structure had to be fixed to the weighed element ratio.

The data of the single crystal structure refinements, in which the M sites are sorted according to increasing lead content (Table 3) and all atomic parameters were transformed to a standardized setting according to STRUCTURE TIDY [24], have been deposited [25].

2.3. Rietveld Refinements

The X-ray powder data used for the full-pattern *Rietveld* refinements were collected on the above mentioned powder diffractometer (Dectris Mythen 1K detector) in the 2Θ range 2 to 41.8° with a step width of 1.17° and an exposure time of 180 s/step. The *Rietveld* refinements were performed using the programs GSAS [26] and EXPGUI [27].

The powder pattern of the sample $SrPb_{0.5}Sn_{2.5}$ (**5p**) could be fully indexed and refined starting from the crystal data of $SrSn_3$ (Mg_3In-type [28]). The obtained lattice parameters (a = 696.50(6), c = 3324.9(5) pm, V/f.u. = 116.4 × 10^6 pm^3), which are somewhat larger than those of the binary compound $SrSn_3$ (a = 694.0, c = 3301.0 pm [3]) already indicated a small lead content. The refinement of the Sn/Pb ratio of the $M(2)$ position (R_P = 0.0611, wR_P = 0.0876) resulted in 28(1)% lead for this site and an overall composition of $SrPb_{0.42}Sn_{2.58}$, which is in good agreement with the weighed sample composition.

For the *Rietveld* refinement of the structure of the phase obtained from the sample $SrPb_{0.95}Sn_{2.05}$ (**6p**), the model of the $PuAl_3$-type, which allowed the full indexing of the powder pattern, was used. The lattice parameters converged at a = 701.06(1) and c = 1683.29(5) pm (V/f.u. = 119.4 × 10^6 pm^3; R_P = 0.0212, wR_P = 0.0270 and R_{F2} = 0.0580). The statistical Sn:Pb ratio of the $M(2)$ position could be refined to 38(1):62(1)%, which is slightly Pb-richer than the relation of the single crystal refinements of **6** ($SrSn_{2.05}Pb_{0.95}$). Accordingly, the lattice parameters are somewhat increased.

2.4. Band Structure Calculations

DFT calculations of the electronic band structure were performed for the border phases of the compound series, $BaSn_3$ and $SrPb_3$, as well as for the two polymorphs of $BaPb_3$ and for $SrSn_3$ using the FP-LAPW method (programs WIEN2K [29] and ELK [30], cf. Table 4). Herein, the exchange-correlation contribution was described by the *Generalized Gradient Approximation* (GGA) of *Perdew, Burke* and *Ernzerhof* [31]. Muffin-tin radii were chosen as 121.7 pm (2.3 a.u.) for all atoms. Cutoff energies used are E_{max}^{pot} = 190 eV (potential) and E_{max}^{wf} = 170 eV (interstitial PW). Further parameters (e.g., number of k points) and selected results of the calculations are collected in Table 4. Electron densities, the *electron localisation function* (ELF) [32,33] and *Fermi* surfaces were visualized using the programs XCRYSDEN [34] and DRAWXTL [35]. A *Bader* 'atoms in molecules' (AIM) analysis of the electron density map was performed to evaluate the charge distribution between the atoms and the heights and positions of the bond, ring and cage critical points [36] using the program CRITIC2 [37,38]. The total DOS (tDOS) plots of all compounds are discussed in Section 3.5.2. For the two border compounds of the series, $BaSn_3$ and $SrPb_3$, plots of the partial DOS (pDOS) and the band structures themselves are already reported elsewhere [3,10,11,14]. Maps of the ELF of $SrSn_3$ based on LMTO-ASA calculations by *Fässler* et al. [3] are in very good agreement with the ELF obtained herein.

Both the band structures and the ELF of the binary compounds reported in the literature are considered in the discussion on the chemical bonding in Section 3.5.2.

Table 4. Details of the calculation of the electronic structures of BaSn$_3$, BaPb$_3$ (BaPb$_3$- and Mg$_3$In-type), SrSn$_3$ and SrPb$_3$. (r_{MT}: Muffin tin radius; K_{max}: maximal wavevector for the PW in the interstitium; BCP: bond critical point; RCP: ring critical point; IBZ: irreducible part of the *Brillouin* zone; V_{BB}: volume of the *Bader* basins).

Compound		**BaSn$_3$**	**BaPb$_3$**		**SrSn$_3$**	**SrPb$_3$**
Structure type		Ni$_3$Sn	BaPb$_3$	Mg$_3$In	Mg$_3$In	Cu$_3$Au
Crystal data		[5]	[11]	Tables 2 and 3	[3]	
R_{MT} (all atoms)		⊢		127.0 pm (2.4 a.u.)		⊣
$R_{MT} \cdot K_{max}$		⊢		8.0		⊣
k-points/BZ		820	1000	⊢ 1000	⊣	1295
k-points/IBZ		72	110	⊢ 110	⊣	75
Monkhorst-Pack-Grid		9 × 9 × 10	10 × 10 × 10	⊢ 10 × 10 × 10	⊣	10 × 10 × 9
DOS		⊢		Figure 12		⊣
Band structure		[3,10,11]			[3]	[10,11]
Bonds:	label					
Electron	a	0.237 (305.9)	0.204 (319.3)	0.218 (315.0)	0.245 (302.6)	0.125 (351.4)
density	b	0.171 (326.7)	0.143 (343.7)	0.138 (346.4)	0.168 (339.2)	
at BCP/RCP(r)	c		0.138 (345.3)	0.136 (346.6)	0.147 (329.2)	
[$e^- 10^{-6}$ pm^{-3}]	d		0.109 (364.5)	0.133 (347.8)	0.159 (334.6)	
(d [pm])	e			0.131 (350.7)	0.150 (337.4)	
	f			0.093 (378.0)	0.116 (359.4)	
	r	0.190	0.152	0.164	0.198	0.088
Atoms:	label					
Charge	Sr/Ba(1)	+1.137 (28.6)	+1.093 (28.6)	+1.071 (28.5)	+1.255 (20.1)	+1.233 (20.3)
distribution	Sr/Ba(2)		+1.067 (28.4)	+1.069 (28.2)	+1.249 (19.8)	
after *Bader*	Sn/Pb(1)	−0.379 (32.1)	−0.359 (34.0)	−0.370 (33.9)	−0.440 (31.5)	−0.418 (34.5)
(V_{BB} [10^6 pm^3])	Sn/Pb(2)		−0.358 (34.5)	−0.344 (34.5)	−0.395 (31.6)	

3. Results and Discussion

3.1. Description of the Crystal Structures and Their Respective Phase Widths

The structures of all tri-tetrelides exhibit hexagonal close packed ordered AM_3 layers containing M kagomé (3.6.3.6.) nets. Depending on the stacking sequence, the M polyanion resembles the oxygen substructure of the hexagonal [face-sharing octahedra, |:AB:| ≡ h stacking, Ni$_3$Sn-type, border compound BaSn$_3$, Figure 1a] or the cubic [corner-sharing octahedra, |:ABC:| ≡ c stacking, Cu$_3$Au-type, border phase SrPb$_3$, Figure 1f] perovskite. In between these border phases, a series of structures occur, in which the amount of cubic stacking increases continuously and simultaneously the size of the blocks of face-sharing octahedra decreases.

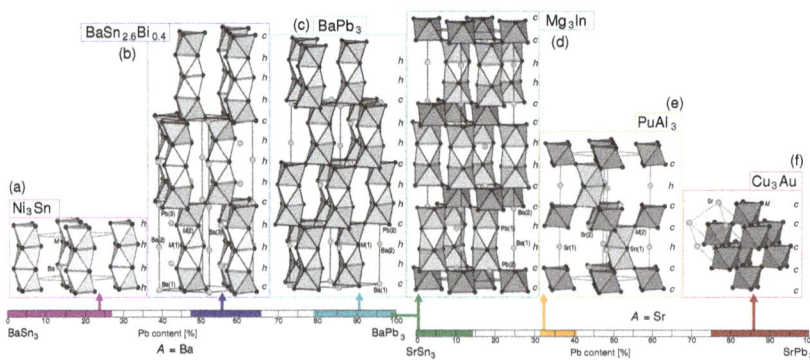

Figure 1. Crystal structures and phase widths (coloured bars) of the different stacking sequences observed in barium (left) and strontium (right) tri-stannides/plumbides (cf. the Figures 2–7 for enlarged views supplemented by the labeling of interatomic distances).

3.1.1. Ni_3Sn-Type Compounds

In the long-known binary stannide $BaSn_3$ [4,5], which forms the hexagonal Ni_3Sn-type ($hP8$, $P6_3/mmc$), up to 28% of the Sn atoms can be substituted against Pb (magenta bar in Figure 1). The crystal structure of the lead-richest phase $BaSn_{2.16}Pb_{0.84}$ (**1**) is depicted in Figure 2 in a polyhedra representation. The interatomic distances of this mixed stannide/plumbide are collected in Table 5. The pure h stacking of the hexagonal planar layers $[BaM_3]$ results in infinite columns of $[M_{6/2}]$ octahedra sharing two opposite faces, which are running along the c axis.

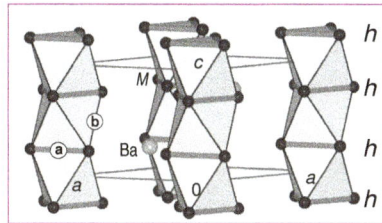

Figure 2. Crystal structure of $BaSn_{2.16}Pb_{0.84}$, **1** (Ni_3Sn-type; bond distances cf. Table 5 [35]).

Table 5. Selected interatomic distances (pm) in the crystal structure of $BaSn_{2.16}Pb_{0.84}$, **1** (Ni_3Sn-type). (cf. Figure 2 for the distance labels).

Atoms		Distance	freq.	CN		Atoms		Distance	lbl.	freq.	CN
Ba	$-M$	364.2(1)	6×			M	$-M$	313.6(2)	a	2×	
	$-M$	366.3(1)	6×	12			$-M$	332.0(1)	b	4×	
							$-Ba$	364.2(1)		2×	
							$-Ba$	366.3(1)		2×	10
							$-M$	412.6(2)		2×	+2

$[M_3]$ three-membered rings with short M–M bonds of length 313.6 pm (label **a**) form the common faces of the $[M_6]$ octahedra of the columns. The remaining octahedra edges are with 332.0 pm (**b**) significantly larger. In the planar M kagomé (3.6.3.6.) nets of the $[BaM_3]$ hexagonal layers, the two remaining M–M contacts are strongly elongated and lie with a distance of 412.6 pm clearly outside any bonding range. Thus, the $[M_3]$ triangles of the 3.6.3.6. net are alternatingly compressed and expanded. Therewith, the M coordination is reduced from the anticuboctahedral 12 (8 M and 4 Ba) surrounding in the ideal hexagonal close packing to an overall 10 (6 + 4) coordination. Compared to the binary border phase $BaSn_3$ [4], the M–M bond lengths (e.g., d^a_{Sn-Sn} = 305.3 pm) and the unit cell volume (cf. discussion in Section 3.3) are increased as expected. The barium cations are coordinated by 12 Sn/Pb atoms in a next to regular anticuboctahedral fashion, with Ba–M distances of 364.2 and 366.3 pm.

3.1.2. $BaSn_{2.6}Bi_{0.4}$-Type Compounds

An increased lead content of 47 to 66% (dark blue bar in Figure 1) causes the formation of the hexagonal $BaSn_{2.6}Bi_{0.4}$-type ($hP32$, $P6_3/mmc$, [18]) with a $(hhhc)_2$ stacking and blocks of four face-sharing octahedra (Figure 3). Selected interatomic distances of both border phases of this type, $BaSn_{1.58}Pb_{1.42}$ (**2**) and $BaSn_{1.01}Pb_{1.99}$ (**3**), are collected in Table 6.

The structure type contains both three crystallographically different barium and three M sites (Table 3). The latter are unequally taken by Sn and Pb: The $M(3)$ position, which forms the $[Ba(1) Pb(3)_3]$ layers of the c stacking is occupied by lead exclusively. Reversely, the $M(1)$ position of a h stacked layer with pure h neighbor layers shows the lowest lead content (9.5–23.7%). The intermediate $M(2)$ position of the hexagonal stacked $[BaM(2)_3]$ layer with an adjacent h and c layer contains an intermediate amount of lead (40.1–70.6%). According to the stacking, the $M(1)$ and $M(2)$ atoms of the *h* layers show a 6 + 4 coordination and form kagomé nets with two different triangles, whereas the lead atoms Pb(3) exhibit the ideal 8 + 4 surrounding and form an ideal 3.6.3.6. net among each other. The [*M*(1)$_3$] three-membered rings forming the common faces between face-sharing octahedra contain the shortest *M*–*M* bonds within the structure (d^a = 302.8–306.3 pm), whereas the analogous [*M*(2)$_3$] ring is enlarged due to the increased Pb proportion (d^c = 313.3–318.5 pm). All Ba cations are coordinated by 12 *M* atoms in a cuboctahedral [Ba(1) and Ba(3)] or anticuboctahedral [Ba(2)] manner. The Ba–*M* distances are all lying in the narrow range 363.4 to 369.4 pm.

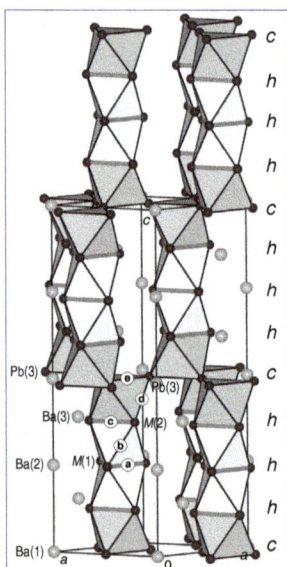

Figure 3. Crystal structure of BaSn$_{1.58}$Pb$_{1.42}$, **2** (BaSn$_{2.6}$Bi$_{0.4}$-type; bond distances cf. Table 6 [35]).

Table 6. Selected interatomic distances (pm) in the crystal structures of the two border compounds BaSn$_{1.58}$Pb$_{1.42}$ (**2**) and BaSn$_{1.01}$Pb$_{1.99}$ (**3**) of the BaSn$_{2.6}$Bi$_{0.4}$-type (cf. Figure 3 for the distance labels).

Atoms		Distances in		freq.	CN	Atoms		Distances in		lbl.	freq.	CN
		2	**3**					**2**	**3**			
Ba(1)	–Pb(3)	363.4(1)	364.1(1)	6×		*M*(1)	–*M*(1)	302.8(3)	306.3(2)	a	2×	
	–*M*(2)	366.6(1)	366.5(1)	6×	12		–*M*(2)	332.4(1)	336.2(1)	b	4×	
							–Ba(3)	354.5(2)	356.3(1)		2×	
							–Ba(2)	365.1(1)	365.6(1)		2×	6 + 4
Ba(2)	–*M*(1)	365.1(1)	365.6(2)	6×								
	–*M*(2)	368.5(1)	369.4(1)	6×	12	*M*(2)	–*M*(2)	313.3(2)	318.5(1)	c	2×	
							–*M*(1)	332.4(1)	336.2(1)	b	2×	
Ba(3)	–*M*(1)	354.5(2)	356.3(1)	3×			–Pb(3)	340.9(1)	343.1(1)	d	2×	
	–*M*(2)	365.4(1)	365.8(1)	6×			–Ba(3)	365.4(1)	365.8(1)		2×	
	–Pb(3)	368.2(2)	369.3(1)	3×	12		–Ba(1)	366.6(1)	366.5(1)			
							–Ba(2)	368.5(1)	369.4(1)			6 + 4
						Pb(3)	–*M*(2)	340.9(1)	343.1(1)	d	4×	
							–Pb(3)	363.4(1)	364.1(1)	e	4×	
							–Ba(1)	363.4(1)	364.1(1)		2×	
							–Ba(3)	368.2(2)	369.3(1)		2×	8 + 4

3.1.3. BaPb$_3$-Type Compounds

The stability range of the rhombohedral BaPb$_3$-type sequence $(hhc)_3$ $(hR36, R\bar{3}m,$ Figure 4) starts at a lead proportion of 78% and reaches up to the pure plumbide BaPb$_3$ [9] (cyan bar in Figure 1). This structure type, which should be termed BaPb$_3$-type in consideration of the first structure determination in 1964 [9], is sometimes also denoted as TaCo$_3$-type. The interatomic distances of the Sn-richest border phase BaSn$_{0.65}$Pb$_{2.35}$ (**4**) are summarized in Table 7.

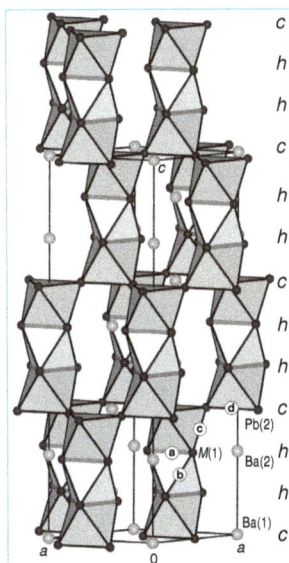

Figure 4. Crystal structure of BaSn$_{0.65}$Pb$_{2.35}$, **4** (BaPb$_3$-type; bond distances cf. Table 7 [35]).

Table 7. Selected interatomic distances (pm) in the crystal structure of BaSn$_{0.65}$Pb$_{2.35}$ (**4**) forming the BaPb$_3$-type (cf. Figure 4 for the distance labels).

Atoms		Distance	freq.	CN	Atoms		Distance	lbl.	freq.	CN
Ba(1)	−M(1)	360.0(2)	3×		M(1)	−M(1)	314.2(3)	a	2×	
	−Pb(2)	364.8(2)	3×			−M(1)	339.5(2)	b	2×	
	−M(1)	366.0(1)	6×	12		−Pb(2)	342.6(1)	c	2×	
						−Ba(1)	360.0(2)			
Ba(2)	−Pb(2)	364.4(1)	6×			−Ba(1)	366.0(1)		2×	
	−M(1)	368.3(1)	6×	12		−Ba(2)	368.3(1)			6 + 4
					Pb(2)	−M(1)	342.6(1)	c	4×	
						−Pb(2)	364.4(1)	d	4×	
						−Ba(2)	364.4(1)		2×	
						−Ba(1)	364.8(2)		2×	8 + 4

In the lead-rich compounds of the Ba series, the planar nets [BaM_3] follow the rhombohedral $(hhc)_3$ stacking sequence and the blocks of face-sharing octahedra are now consisting of three octahedra only (Figure 4). Within these trimers, the [M_6] octahedra share common triangular faces [$M(1)_3$] with the shortest interatomic distances of the whole structure (d^a = 314.2 pm, cf. Table 7). As expected from the tin content, this and all further $M-M$ distances are somewhat shortened with respect to the isotypic pure plumbide (e.g., d^a = 319.3 pm, [11]). The octahedra trimers are connected via the six corners formed by the pure lead position [Pb(2)]. According to the point group symmetry $.\frac{2}{m}$ of the

Pb(2) position, these atoms form ideal kagomé nets at $z = 0$, $\frac{1}{3}$ and $\frac{2}{3}$ in the unit cell. The mesh width of this net, i.e., the Pb(2)–Pb(2) distances (4×) are comparatively large 364.4 pm, whereas the remaining M–M distances are of intermediate lengths (339.5 and 342.6 pm). Again, the atoms forming the h layers exhibit a 6 + 4 coordination, whereas the c net position [Pb(2)] has the full 8 M + 4 Ba surrounding of a dense packing. The Ba(1) cations show a cuboctahedral, Ba(2) an anticuboctahedral, both fairly regular ($d_{Ba-M} = 360.0$–368.3 pm), twelvefold M coordination.

3.1.4. Mg$_3$In-Type Compounds

A second new polymorph of BaPb$_3$ (denoted BaPb$_3$-II hereafter) crystallizes in the Mg$_3$In-type structure with a further increased amount of cubic sequences [$(hhcc)_3$; $hR48$, Figure 5]. This form of BaPb$_3$ is thus isotypic with the border phase SrSn$_3$ [3] of the respective Sr series and with the lead-containing compounds up to a Pb content of 14% (SrSn$_{2.58}$Pb$_{0.42}$, **5p**; green bar in Figure 1). The structure model of these compounds should be denoted Mg$_3$In-type, because the crystallographic data of this particular compound were the first one published in 1963 [28]. Two years later, an isotypic structure was reported for the high-temperature form of PuGa$_3$ and only much later (2000, [3]) the tri-stannide SrSn$_3$ was added. The two latter compounds are nevertheless widely (e.g., in the *Inorganic Crystal Structure Database* [39]) used to address this structure type.

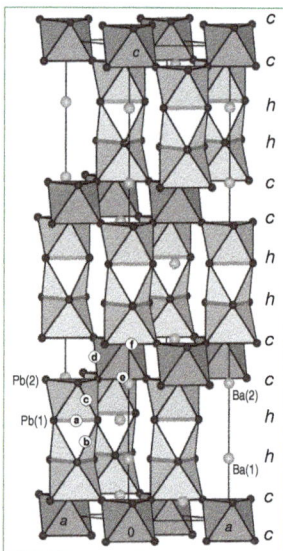

Figure 5. Crystal structure of BaPb$_3$-II, **5** (Mg$_3$In-type; bond distances cf. Table 8 [35]).

In the structure of BaPb$_3$-II (Figure 5) the rhombohedral stacking sequence $(hhcc)_3$ results in octahedra trimers, which are not directly connected via corners as in BaPb$_3$-I, but an additional layer of corner-connected [Pb(2)$_6$] octahedra is interspersed. Overall, a 1:1 relation of h and c layers, i.e., Pb(1) and Pb(2) atoms, are reached. In addition, in this case, the Pb–Pb distances of the common [Pb(1)$_3$] triangles of the face-sharing lead octahedra are with 315.0 pm the shortest bonds within the structure. All other Pb–Pb distances are found in the range 346–378 pm, whereby the largest values are again in the c-stacked kagomé nets [Pb(2)–Pb(2)]. In contrast to the c layers in the BaPb$_3$-type structures, the 3.6.3.6. Pb(2) nets exhibit two different triangles with Pb(2)–Pb(2) edges of 350.7 and 378.0 pm. Nevertheless, the [Pb(2)$_3$] triangles in the h layers are much more different in size (315.0/413.7 pm). Due to these compressed/expanded triangles, the Pb(1) atoms of the common octahedra faces are surrounded by 6 Pb and 4 Ba cations only, whereas Pb(2) of the c net exhibits the complete 12-fold coordination, even though two lead neighbors are already at a slightly larger distance than the four Ba cations (Table 8). The cations themselves again show the ideal 12-fold lead coordination, with 50% cuboctahedral [Ba(2)] and 50% anticuboctahedral [Ba(1)] arrangement. The Mg_3In-type structure has been already observed (and predicted by *van Vucht* in 1966 [7]) for pure plumbides with a small substitution of barium by calcium or strontium (for $A_xBa_{1-x}Pb_3$: $x = 0.03$ for Ca and 0.10 for Sr [11]). It also occurs in the compound series $Ba(Sn_xBi_{1-x})_3$ [18] and from $SrSn_3$ to $SrSn_{2.58}Pb_{0.42}$ (**5p**). Unfortunately (due to problems with the crystal quality in the Sr series, cf. Section 2.2) the structure of the latter compound has been refined from powder data only (Section 2.3). As expected, the lattice parameters and all $M–M$ bonds are slightly elongated compared to $SrSn_3$ and the lead atoms are substituted at the c stacked layers, i.e., the $M(2)$ position.

Table 8. Selected interatomic distances (pm) in the crystal structure of $BaPb_3$ (Mg_3In-type structure). (cf. Figure 5 for the distance labels).

Atoms		Distance	freq.	CN	Atoms		Distance	lbl.	freq.	CN
Ba(1)	–Pb(1)	361.7(2)	3×		Pb(1)	–Pb(1)	315.0(2)	a	2×	
	–Pb(2)	365.6(2)	3×			–Pb(1)	346.4(2)	b	2×	
	–Pb(1)	366.2(1)	6×	12		–Pb(2)	346.6(1)	c	2×	
						–Ba(2)	361.6(2)			
Ba(2)	–Pb(1)	361.6(2)	3×			–Ba(1)	361.7(2)			
	–Pb(2)	364.6(1)	6×			–Ba(1)	366.2(1)		2×	6 + 4
	–Pb(2)	365.0(2)	3×	12	Pb(2)	–Pb(1)	346.6(1)	c	2×	
						–Pb(2)	347.8(2)	d	2×	
						–Pb(2)	350.7(2)	e	2×	
						–Ba(2)	364.6(1)		2×	
						–Ba(2)	365.0(2)			
						–Ba(1)	365.6(2)			
						–Pb(2)	378.0(2)	f	2×	6 + 4 + 2

In the vast series of tri-metallides of the structure family considered, polymorphism (like the one observed herein for $BaPb_3$) is very frequent. An instructive example are the tri-gallides of the rare earth elements, were the smaller Er, Tm and Lu cations form Cu_3Au-type structures, whereas Tb, Dy and Ho are di- or even trimorphic [40]. For the two modifications of $BaPb_3$, the difference of the unit cell volumes per formula unit ($V/\text{f.u.}$) is even smaller than 0.1% [131.16 (I) compared to 131.07×10^6 pm^3 (II)] and an experimental hint towards the phase relation between the two polymorphs is also not obvious.

3.1.5. $PuAl_3$-Type Compounds

In the Sr series, a Pb content of 32% ($SrSn_{2.05}Pb_{0.95}$, **6**) to 41% ($SrSn_{1.77}Pb_{1.23}$, **6p**) causes a small existence region of the hexagonal $PuAl_3$-type ($hP24$, $P6_3/mmc$ [41]) with a $(hcc)_2$ stacking (golden bar in Figure 1). The structure and selected bond lengths are found in Figure 6 and Table 9.

The increased proportion of c stacking in the $PuAl_3$-type reduces the size of the blocks of face-sharing octahedra to dimers, which are separated by additional c stacked [$M(2)6$] corner-sharing octahedra. Again, tin prefer the occupation of the h layer and the shortest distances are found in the common face triangle inside this layer ($d^a_{Sn(1)–Sn(1)} = 302.9$ pm). The triangles of the h net are thus of very different size, whereas the $M(2)$ kagomé nets of the c layers are only slightly distorted (345.0, **d**/352.0, **e** pm). The overall coordination of the h and c layer M atoms and the strontium cations again fits the situation throughout the whole structure family.

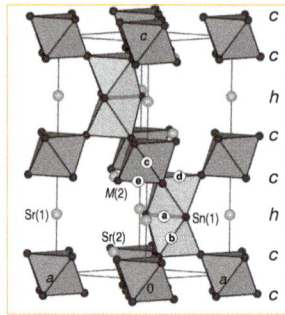

Figure 6. Crystal structure of $SrSn_{2.05}Pb_{0.95}$, **6** ($PuAl_3$-type; bond distances cf. Table 9 [35]).

Table 9. Selected interatomic distances (pm) in the crystal structure of $SrSn_{2.05}Pb_{0.95}$ ($PuAl_3$-type). (cf. Figure 6 for the distance labels).

Atoms		Distance	freq.	CN	Atoms		Distance	lbl.	freq.	CN
Sr(1)	–M(2)	348.4(2)	6×		Sn(1)	–Sn(1)	302.9(8)	a	2×	
	–Sn(1)	349.5(1)	6×	12		–M(2)	339.9(2)	b	4×	
						–Sr(2)	347.8(6)		2×	
Sr(2)	–Sn(1)	347.8(6)	3×			–Sr(1)	349.5(1)		2×	6 + 4
	–M(2)	349.1(1)	6×			–Sn(1)	394.1(8)		2×	
	–M(2)	353.2(7)	3×	12						
					M(2)	–M(2)	339.3(4)	c	2×	
						–Sn(1)	339.9(2)	b	2×	
						–M(2)	345.0(4)	d	2×	
						–Sr(1)	348.4(2)			
						–Sr(2)	349.1(1)		2×	
						–M(2)	352.0(4)	e	2×	
						–Sr(2)	353.2(7)			8 + 4

3.1.6. Cu_3Au-Type Compounds

The strontium series is terminated by the pure $(c)_3$ stacking of $SrPb_3$ (Cu_3Au-type, *cP4*, *Pm3m*), which starts at a lead content of approximately 75%. Because of problems with the single crystal quality, a reliable refinement of the Sn/Pb occupation was not possible, so that some uncertainty of the exact composition remains. The very broad reflections of the powder diffraction pattern prohibit even a full-pattern *Rietveld* refinement and a decision is not possible, whether the tetragonal distortion of the cubic Cu_3Au-type, which is reported in the literature for the binary tri-plumbide $SrPb_3$ [6], also arises in the ternary tin-containing phase.

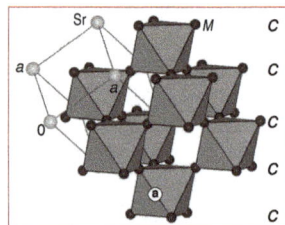

Figure 7. Crystal structure of $SrSn_{0.75}Pb_{2.25}$, **7** (Cu_3Au-type; bond distances cf. Table 10 [35]).

In accordance with the presence of smaller Sn atoms, the lattice parameter and therewith the *M*–*M* distance of 350.3 pm is somewhat decreased compared to the two distances found in the tetragonal border phase $SrPb_3$ (d_{Pb-Pb} = 351.1, 353.2 pm [6]). The structure terminates the series with a pure *c* stacking and a polyanion, which consists of ideal corner-sharing $[M_{6/2}]$ octahedra. All atoms exhibit an ideal cuboctahedral surrounding with one unique distance of 350.3 pm (cf. Table 10).

Table 10. Selected interatomic distances (pm) in the crystal structure of $SrSn_{0.75}Pb_{2.25}$ (Cu_3Au-type structure).

Atoms		Distance	freq.	CN		Atoms		Distance	lbl.	freq.	CN
Sr	−*M*	350.3(1)	12×	12		*M*	−*M*	350.3(1)	a	6×	
							−Sr	350.3(1)		6×	6 + 6

3.2. Stacking Sequences: Geometric Aspects

The series of mixed Sn/Pb tri-tetrelides discussed above, as well as similar systematic studies on mixed Ca/Sr/Ba stannides [10] and plumbides [11], clearly show, that the amount of *h*/*c* stacking in the structure family of AB_3 superstructures of dense sphere packing and hence the sequence of the different structure types is essentially determined by the radius ratio of the two elements. This becomes apparent from the coherent progress of the amount of hexagonal stacking with the radius ratio of the two elements shown in Figure 8. Similar plots are presented for a great variety of compounds in the literature [7,8]. In contrast to these works, in which metallic radii were used to estimate the atomic sizes, ionic radii (after *Shannon* [42], for CN = 12) were used herein to estimate the size of the *A* cations, and metallic radii (after *Gschneidner* [43], for CN = 12) were only applied to quantify the size of the Sn and Pb atoms. This choice of radii is in agreement with the next to complete electron transfer from the alkaline-earth elements *A* to Sn/Pb (cf. Section 3.5.2). The diagram shown in Figure 8 demonstrates the direct relation between the stacking sequence and the radius ratio, independent of the type of the *M* atom, i.e., both for Sn (squares and thin dashed line), Pb (circles and thin solid line) or Sn and Pb (with a distinct distribution at the *h*/*c* layer, colored triangles and bold lines).

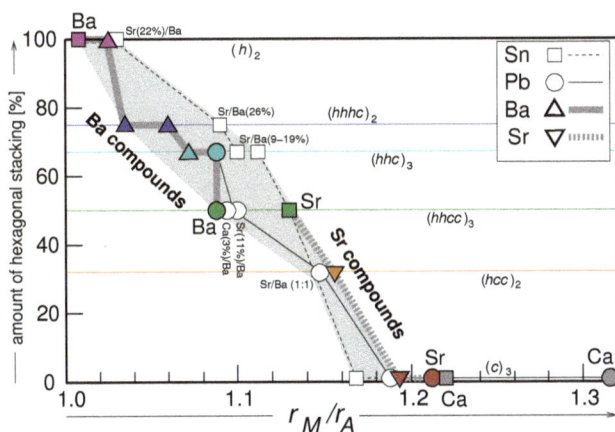

Figure 8. Amount of hexagonal stacking plotted versus the radius ratio r_M:r_A in (mixed) alkaline-earth tri-stannides (thin squares and dashed line) and -plumbides (thin circles and line) and the mixed stannides/plumbides (triangles and bold gray lines) reported in this work.

This general behaviour can be explained by the decreased (10) coordination number of the atoms of the *h* layers ($6 \times M + 4 \times A$) compared to the ideal twelvefold ($8 \times M + 4 \times A$) coordination of the *M* atoms of *c* stacked layers (Figure 9).

However, the reason for the reduction of the coordination sphere of h-type atoms from 12 to 10 is the compression of the common $[M_3]$ triangle of two face-sharing octahedra, which is causally linked to covalent bonding contributions within the M polyanion (cf. Section 3.5).

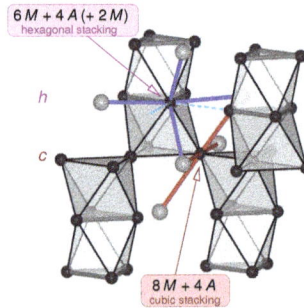

Figure 9. Comparison of the characteristic coordination spheres of Sn/Pb atoms of a hexagonal (violet-blue) and a cubic (red) stacked layer [35].

In electron poorer compounds (like for example $CaHg_3$ or $SrHg_3$ [44]), which crystallize in the Ni_3Sn-type, the M–M triangles do not show shortened edges/bonds (cf. also the discussion on the value of the atomic x parameter in the Ni_3Sn and related structures by *Havinga* in [8]).

3.3. Molar Volumes and Variation of the c/a Ratio

Regardless of the different stacking sequences, the molar volumes of all tri-tetrelides follow *Vegards* rule (gray circles and thin black lines in Figure 10). The normalized interlayer distance, i.e., the c/a ratio/two layers (open circles and gray lines) decreases continuously with (i) increasing h stacking (thick gray lines) and (ii) with the Sn content of each distinct structure type. This increasing hexagonal unit cell basis is a consequence of the covalent bonding contributions and the stereochemically active lone electron pair (l.e.p., see Section 3.5) of the tin (and less pronounced the lead) atoms of the h layers in the electron-richer compounds of the AB_3 structure family. Simple packing/geometric arguments failed to explain this general trend, which is similarly observed for other electron-richer compounds of this structure family [7,8].

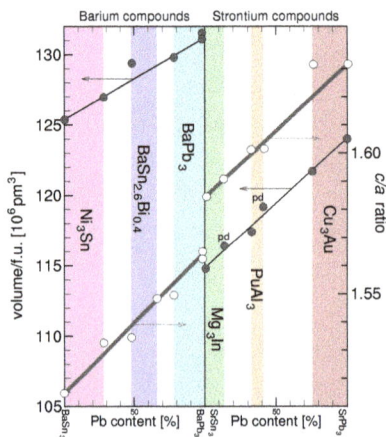

Figure 10. Normalized unit cell volumes V/f.u. and c/a ratios (for two layers) of the mixed tri-stannides/plumbides of barium (left) and strontium (right).

3.4. Sn/Pb Distribution: 'Coloring' Aspects

Concerning the 'coloring' [19], i.e., the preference of tin and lead to occupy the differently stacked layers, the clear preference of Sn atoms to occupy the *h* nets is valid throughout the whole family of the tri-tetrelides. Arguments for this very uniform coloring are at one hand the reduced coordination sphere of the *h* layer atoms (cf. Section 3.2), which may better fit the smaller *M* atom types. On the other hand, the larger Sn–Sn (compared to Pb–Pb) single bond energies direct tin to take those sites in the structure, where generally the shortest bonds are formed; this is unequivocally the *h* position.

3.5. Chemical Bonding, Electronic Structure

Calculations of the electronic band structure, based on different levels of theory, are already reported in vast detail for several border phases of the title compound series. The results of LMTO-ASA calculations of $SrSn_3$ [3] and $BaSn_3$ [5] are in very good agreement with the FP-LAPW results discussed in [10,11]. The obtained electron densities of these DFT calculations reveal the different types of Sn–Sn bonds as well as the presence of the l.e.p. at the *h*-stacked tin atoms [10]. The related ELF exhibits comparatively low (approx. 0.6) maxima at the bonds of the common octahedra edges and for the l.e.p. of *h*-type atoms [3]; whereas the ELF obtained from EH calculations shows significantly more pronounced features (e.g., [5] for $BaSn_3$). The band structures of the cubic phases like $CaPb_3$ and $SrPb_3$ were also calculated using FP-LMTO [14] and FP-LAPW [10,11] methods. A detailed discussion of the band structure from the point of view of chemical bonding can be found in [10]. In this work, the literature's theoretical work is complemented by FP-LAPW calculations of the new polymorph of $BaPb_3$, which—according to its isotypy to $SrSn_3$—allows for a direct comparison of an Sn with an isotypic (Mg_3In-type) Pb compound. Corresponding calculations of the border phases $BaSn_3$, $BaPb_3$-I, $SrSn_3$ and $BaPb_3$ (Table 4) are also included for comparison and for a detailed analysis of the electron density. Independent of the stacking of the close-packed layers, the total density of states (tDOS) show pronounced minima next to the *Fermi* level (14 v.e./f.u.), which implies the validity of simple electron counting rules.

3.5.1. Electron Count (*Zintl*/*Wade*/*mno*)

Compounds with the pure *h* stacking like e.g., $BaSn_3$ exhibit three-membered rings with short *M*–*M* bonds, which are only slightly larger than single bond distances, and the calculated electron density shows the highest electron density at the respective bond critical points ($\rho_{BCP} = 0.237$ e$^-10^{-6}$ pm^{-3}, Table 4 and [10]). Neglecting all further *M*–*M* contacts—as proposed in [5]—the compound contains *Zintl* anions $[M_3]^{2-}$ with an aromatic π system of two electrons. Admittedly, the results of the band structure calculations [10] show, that this simplified model is only of limited applicability: No (pseudo) band gap can be found at the *Fermi* level (cf. Figure 12) and the number of p_z-type bands below E_F is not compatible with an aromatic π system. However, the evident stability of the whole structure family at (or close to) 14 v.e. per formula unit (i.e., $4\frac{2}{3}$ v.e./*M*) can be explained both for the *h* and the *c* stacking (and therewith also for all stacking variants in between) when taking the 'polyaromatic' bonding character (cf. the longer *M*–*M* bonds and the ring critical points, Section 3.5.2) into account and extending the *Zintl* concept by *Wade*'s electron counting rules [45] or the somewhat more comprehensive *mno* rule of *Jemmis* [46]:

c   The bonding within *c* stacked compounds like e.g., $SrPb_3$, in which $[Pb_6]$ octahedra are fused via common corners (Figure 7), can be explained by a direct comparison with the electron-precise boride CaB_6 containing also octahedral *closo* clusters B_6^{2-} (20 v.e.), which are in this case connected via 2e2c bonds (*exo*-bonds). Subtracting the six v.e. needed for these *exo*-bonds, each boron atom contributes $2\frac{1}{3}$ v.e. to stabilize one octahedral *closo* cluster. In tri-tetrelides two octahedra are directly fused via corners, i.e., each Pb atom participates in two *closo* octahedral clusters. Thus, the polyanion needs $2 \times 2\frac{1}{3} = 4\frac{2}{3}$ v.e./Pb to obey *Wade*'s rule. This corresponds to the observed v.e. number ($2SrPb_3 \longrightarrow 2Sr^{2+} + Pb_6^{4-}$: $6 \times 4 + 4$ v.e./f.u. $\equiv 4\frac{2}{3}$ v.e./Pb). and to the fact, that the tDOS exhibits a very pronounced minimum close to the *Fermi* level, at 13.9 v.e./f.u. (bottom of Figure 12, black arrow).

h   The *mno* rules and the similarities to the 'polyaromatic' bonding in borides and molecular boranes/boranates can be also used to explain the electron count of the *h* stacked atoms: Figure 11 shows the formal splitting of the v.e. in the boranate $[B_9H_6]^-$, which satisfies the *mno* rules [46] $(9 + 2 + 0 = 11$ s.e.p. (s.e.p.: skeleton electron pairs): $\underbrace{9 \times 3}_{B} + \underbrace{6 \times 1}_{H} + \underbrace{1}_{charge} - \underbrace{6 \times 2}_{exo\text{-}BH} = 22$ s.e. $= 11$ s.e.p.). For the formal splitting of these overall number of s.e. to individual B atoms, the six BH atoms contribute 14 s.e. $(6 \times 2\frac{1}{3}$, see above). The remaining 8 s.e. result in $2\frac{2}{3}$ v.e. for each of the three central (*h* stacked) B atoms $(1\frac{1}{3}$ s.e./octahedron). Transfered to the electron balance of $BaSn_3$, where each of the Sn atoms carries an additional lone-electron pair (s^2 electrons), we expect $2 + 2\frac{2}{3} = 4\frac{2}{3}$ v.e./Sn (i.e., 14 v.e./f.u.) to result in a stable cluster.

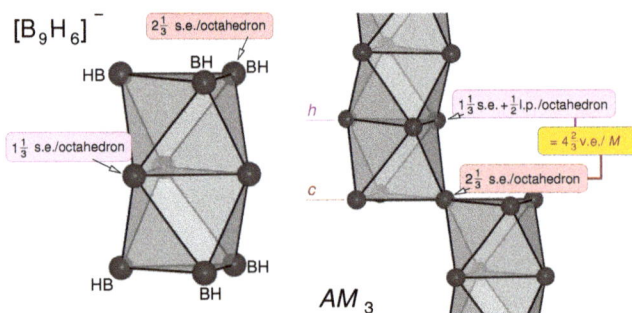

Figure 11. Explanation of the electron count proving the validity of the *Wade-Jemmis mno* rules for the hexagonal and cubic stacking in alkaline-earth tri-tetrelides.

The uniform ideal v.e. number for both the *h* and *c* stacked *M* atoms of $4\frac{2}{3}$ v.e./*M* explains the stability of the polyanion in all tri-tetrelides AM_3, i.e., for compounds with 14 v.e./f.u.

3.5.2. Bandstructure Calculations (DOS, Electron Densities)

In accordance with the electron count discussed above, the calculated total densities of states (tDOS) of all tri-tetrelides (cf. Figure 12 and [10,11]) exhibit distinct minima close to 14 v.e., slightly below the *Fermi* level. The *M* pDOS of *c* and *h* layers are plotted in the familiar color code, magenta arrows indicate the position of the lone-electron pairs (l.e.p.) of the *s*-states of *h*-stacked *M* atoms. The pDOS of the *M* atoms of the polyanion are separated into overall non-bonding *M-s* ($-10...-4$ eV, 6 v.e./f.u.) and bonding *M-p* states (>-4 eV, 8 v.e./f.u.). As expected, this separation increases from the stannides to the plumbides. The low Sr/Ba pDOS below E_F (black lines in Figure 12) is in accordance with the positive *Bader* charges (*q*) of the *A* cations and the pronounced electron transfer from the alkaline-earth element towards the *p* block element. The cation charges vary slightly between $+1.07$ ($BaPb_3$) and $+1.26$ ($SrSn_3$) (Table 4). In mixed *h*/*c* stacked compounds, the different *M* atoms are comparably negatively charged, which is in agreement with the equal formal electron demand and the great importance of geometric parameters for the stacking sequences and the polyanion 'coloring'. Overall, the *M* charges vary between -0.34 and -0.42.

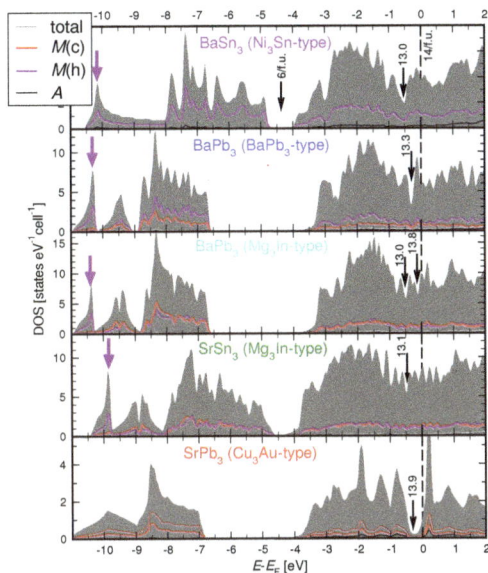

Figure 12. Density of states of the calculated alkaline-earth tri-tetrelides (in eV relative to the *Fermi* level; magenta/red: partial DOS of *M* atoms of a *h/c* layer).

The calculated electron density maps (cf. Figure 13 for a cross-section of the *h* layer in BaPb$_3$-II) exhibit bond (BCP) and ring (RCP) critical points of significant height $\rho_{BCP/RCP}$ (>0.1 e$^-$10^{-6} pm^{-3}) for all labelled *M*–*M* bonds. (In the tetrahedra/octahedra interstices ρ drops down to 0.02 e$^-$10^{-6} pm^{-3}). Thereby ρ_{BCP} decrease with increasing *M*–*M* distances (Table 4). The largest values (i.e., strongest bonds) are found for the edges of the face-sharing [M_6] octahedra (label **a** in Figure 13). In the calculated ELF of both the stannides and the plumbides, only this bond (and the l.e.p.) is recognizable by a small maximum of approx. 0.6 (see also [4]). These three-membered rings are centered by distinct ring critical points of heights between 0.15 (Pb compounds, cf. Figure 13) and 0.20 e$^-$10^{-6} pm^{-3} (Sn compounds).

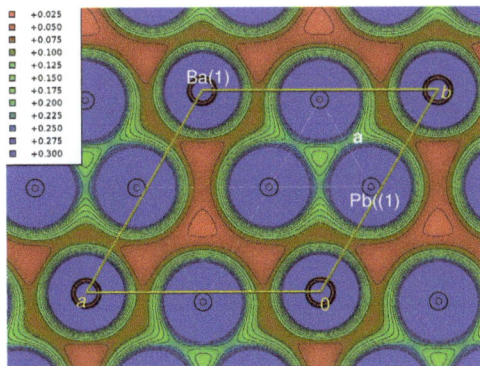

Figure 13. Cross-section of the calculated electron density map [e$^-$10^{-6} pm^{-3}] of the *h* section of BaPb$_3$-II.

The remaining faces of all $[M_6]$ octahedra depicted in the structure drawings above are also centered by RCP with somewhat smaller values of ρ_{RCP}. This and the analysis of the bandstructures of the border compounds $BaSn_3$, $CaPb_3$ and $BaPb_3$ themselves given in the literature [6,10,11,14,15] justify the description of the covalent bonding contributions in the tri-tetrelides as 'polyaromatic', i.e., as analogs of e.g., boron compounds. The $M–M$ distances, which are considerably increased compared to the single bond lengths, are also similar to the larger contacts observed e.g., in the electron-precise *nido* clusters $M_9{}^{4-}$ of several alkali stannides/plumbides [47–52].

Nevertheless, polar compounds of this structure family are known to be electronically somewhat flexible and the minimum of the tDOS generally does not exactly coincide with the *Fermi* level. The comparison of the tDOS of tri-tetrelides with varying h/c stacking ratio displays tendencies towards a small decrease of the tDOS minimum with increasing h-stacked layers (cf. black arrows in Figure 12). A similar trend is obvious in the tin series $CaSn_3/SrSn_3/BaSn_3$, where the tDOS minima are found at 13.9/13.3/13 v.e./f.u. [10]. This shift is in accordance with several experimental observations [7,12,15], e.g., for Tl-substituted plumbides [16] and the series $BaBi_3$–$BaSn_3$ [18]: in both cases, the amount of h stacked layers increases continuously with decreasing v.e. numbers.

4. Summary

The reported systematic experimental, crystallographic and bond-theoretical DFT study of the series of mixed Sn/Pb (M) tri-tetrelides of Sr and Ba (A) AM_3 allows the study of the geometric and electronic criteria, which determine the stacking sequences of planar AM_3 hexagonal layers in the large family of close-packed AB_3 structures. The series of new compounds starts at the binary tri-stannide $BaSn_3$ with a pure hexagonal h_2 stacking ($NiSn_3$-type), in which up to 28% of the Sn atoms can be substituted against Pb. A further increased lead content of 47 to 66% causes the formation of the $BaSn_{2.6}Bi_{0.4}$-type structure with a $(hhhc)_2$ stacking. The stability range of the $BaPb_3$-type sequence $(hhc)_3$ starts at a lead proportion of 78% and reaches up to the pure plumbide $BaPb_3$. A second new polymorph of $BaPb_3$ forms the Mg_3In-type structure with a further increased amount of cubic sequences $[(hhcc)_3]$ and is thus isotypic with the border phase $SrSn_3$ of the respective Sr series. In this series, a Pb content of 32 to 41% creates an existence region of the $PuAl_3$-type with a $(hcc)_2$ stacking. The series is terminated by the pure c stacking of $SrPb_3$ (Cu_3Au-type), which starts at 75% Pb. Even though all compounds are metals, the calculated DOS exhibits pronounced pseudo band gaps slightly below E_F of 14 v.e./f.u. The minimum occurs independent of the type of stacking, so that geometric parameters (radius ratio of the atoms/cations) are the main parameters determining the layer sequence. The application of simple covalent bonding concepts (*Zintl* concept and *Wade/Jemmis mno* rules), which is justified by the distinct bond and ring critical points of the electron density maps, the features of the ELFs and the band structures themselves, allows for the explanation of these DOS minima at approx. 14 v.e./f.u. Nevertheless, metallic (i.e., steep bands crossing E_F) besides the covalent (flat bands near E_F) features are the main characteristic of the band structures of the tri-tetrelides and determine their specific physical properties like e.g., their superconductivity.

Author Contributions: Compound synthesis: M.L. and M.J.; Data collection (powder and single crystal): M.L. and M.J.; Structure refinements: M.L., M.J. and C.R.; Theoretical calculations: M.L., M.J. and C.R.; Writing: M.L., M.J. and C.R.

Acknowledgments: We would like to thank the *Deutsche Forschungsgemeinschaft* for financial support.

Conflicts of Interest: The authors declare no conflict of interest.

References and Notes

1. Fässler, T.F.; Hoffmann, S. Valence compounds at the border to intermetallics: alkali and alkaline earth metal stannides and plumbides. *Z. Kristallogr.* **1999**, *214*, 722–734. [CrossRef]
2. Hume-Rothery, W. Formation of intermetallic compounds. Part IV: The System Calcium-Tin and the compounds $CaSn_3$, $CaSn$ and Ca_2Sn. *J. Inst. Met.* **1926**, *35*, 319–335.

3. Fässler, T.F.; Hoffmann, S. SrSn$_3$–eine supraleitende Legierung mit freien Elektronenpaaren. *Z. Anorg. Allg. Chem.* **2000**, *626*, 106–112. [CrossRef]

4. Ray, K.W.; Thompson, R.G. Study of barium-tin alloys. *Met. Alloys* **1930**, *1*, 314–316.

5. Fässler, T.F.; Kronseder, C. BaSn$_3$, ein Supraleiter im Grenzbereich zwischen Zintl-Phasen und intermetallischen Verbindungen: Realraumanalyse von Bandstrukturen. *Angew. Chem.* **1997**, *109*, 2800–2803. [CrossRef]

6. Damsma, W.; Havinga, E.E. Influence of small lattice deformation on the superconductive critical temperature of alloys with the Cu$_3$Au type structure. *J. Phys. Chem. Solids* **1972**, *34*, 813–816. [CrossRef]

7. Van Vucht, J.H.N. Influence of radius ratio on the structure of intermetallic compounds of the AB$_3$ type. *J. Less Common Met.* **1966**, *11*, 308–322. [CrossRef]

8. Havinga, E.E. Influence of repulsive energy on structural parameters of the close-packed metal structures. *J. Less Common Met.* **1975**, *41*, 241–254. [CrossRef]

9. Sands, D.E.; Wood, D.H.; Ramsey, W.J. The structures of Ba$_5$Pb$_3$, BaPb and BaPb$_3$. *Acta Crystallogr.* **1964**, *17*, 986–989. [CrossRef]

10. Wendorff, M.; Röhr, C. Gemischte Tristannide der Reihe CaSn$_3$–SrSn$_3$–BaSn$_3$: Synthesen, Kristallstrukturen, Chemische Bindung. *Z. Anorg. Allg. Chem.* **2011**, *637*, 1013–1023. [CrossRef]

11. Wendorff, M.; Röhr, C. Gemischte Plumbide (Ca/Sr)$_x$Ba$_{1-x}$Pb$_3$. Strukturchemie und chemische Bindung. *Z. Naturforsch.* **2008**, *63b*, 1383–1394.

12. Havinga, E.E. W-like dependence of critical temperature on number of valence electrons in non-transition metal Cu$_3$Au-type alloys. *Phys. Lett. A* **1968**, *28*, 350–351. [CrossRef]

13. Havinga, E.E.; Damsma, W.; van Maaren, M.H. Oscillatory dependence of superconductive critical temperature on number of valency electrons in Cu$_3$Au type alloys. *J. Phys. Chem. Solids* **1970**, *31*, 2653–2662. [CrossRef]

14. Baranovskiy, A.E.; Grechnev, G.E.; Svechkarev, I.V. Features of the electronic spectrum and anomalous magnetism in the compounds YbPb$_3$, YbSn$_3$, CaPb$_3$ and CaSn$_3$. *Low Temp. Phys.* **2006**, *32*, 849–856. [CrossRef]

15. Baranovskiy, A.E.; Grechnev, G.E.; Mikitik, G.P.; Svechakarev, I.V. Anomalous diamagnetism in the intermetallic compounds CaPb$_3$ and YbPb$_3$. *Low Temp. Phys.* **2003**, *29*, 356–358. [CrossRef]

16. Havinga, E.E.; van Vucht, J.H.N. The crystal structure of Ba(Pb$_{0.8}$Tl$_{0.2}$)$_3$. *Acta Crystallogr.* **1970**, *B26*, 653–655. [CrossRef]

17. Single crystal structure refinement of the strucuture of BaIn$_{0.15}$Pb$_{2.85}$: Hexagonal, space group *P*6$_3$/*mmc*, *a* = 733(1), *c* = 3991(5) pm, *Z* = 14, *R*1 = 0.0494.

18. Ponou, S.; Fässler, T.F.; Kienle, L. Structural complexity in intermetallic alloys: Long-periodic order beyound 10 nm in the system BaSn$_3$/BaBi$_3$. *Angew. Chem.* **2008**, *47*, 3999–4004. [CrossRef] [PubMed]

19. Miller, G.J. The "Coloring Problem" in solids: How it affects structure, composition and properties. *Eur. J. Inorg. Chem.* **1998**, *5*, 523–536. [CrossRef]

20. Yvon, K.; Jeitschko, W.; Parthé, E. *Program LAZY-PULVERIX*; University Geneve: Geneva, Switzerland, 1976.

21. STOE & Cie GmbH. *X-SHAPE (Version 1.03), Crystal Optimization for Numerical Absorption Correction*; STOE & Cie GmbH: Darmstadt, Germany, 2005.

22. Sheldrick, G.M. *SADABS: Program for Absorption Correction for Data from Area Detector Frames*; Bruker Analytical X-ray Systems, Inc.: Madison, WI, USA, 2008.

23. Sheldrick, G.M. A short history of SHELX. *Acta Crystallogr.* **2008**, *A64*, 112–122. [CrossRef] [PubMed]

24. Gelato, L.M.; Parthé, E. STRUCTURE TIDY: A computer program to standardize structure data. *J. Appl. Crystallogr.* **1990**, *A46*, 467–473. [CrossRef]

25. Further details on the crystal structure investigation are available from the Fachinformationszentrum Karlsruhe, Gesellschaft für wissenschaftlich-technische Information mbH, D-76344 Eggenstein-Leopoldshafen 2 on quoting the depository numbers CSD 431093 (BaPb$_3$-II), 431094 (BaSn$_{0.65}$Pb$_{2.35}$) 431095 (BaSn$_{1.01}$Pb$_{1.99}$) 431096 (BaSn$_{1.52}$Pb$_{1.42}$) 431097 (BaSn$_{2.20}$Pb$_{0.80}$) and 431098 (SrSn$_{2.05}$Pb$_{0.95}$), the names of the authors, and citation of the paper (E-mail: crysdata@fiz-karlsruhe.de).

26. Larson, A.C.; Dreele, R.B.V. *General Structure Analysis System (GSAS)*; Los Alamos National Laboratory Report LAUR 86-748; Los Alamos National Laboratory: Los Alamos, NM, USA, 2000.

27. Toby, B.H. EXPGUI, a graphical user interface for GSAS. *J. Appl. Crystallogr.* **2001**, *34*, 210–221. [CrossRef]

28. Schubert, K.; Gauzzi, F.; Frank, K. Kristallstruktur einiger Mg-B3-Phasen. *Z. Metallk.* **1963**, *54*, 422–429.

29. Blaha, P.; Schwarz, K.; Madsen, G.K.H.; Kvasnicka, D.; Luitz, J. *WIEN2K, An Augmented Plane Wave and Local Orbital Program for Calculating Crystal Properties*; TU Wien: Vienna, Austria, 2006; ISBN3-9501031-1-2.

30. Dewhurst, J.K.; Sharma, S.; Nordstrom, L.; Cricchio, F.; Bultmark, F.; Gross, E.K.U. ELK (Vers. 2.1.15), The Elk–FP-LAPW Code. 2013. Available online: http://elk.sourceforge.net (accessed on 5 May 2018).

31. Perdew, J.P.; Burke, S.; Ernzerhof, M. Generalized gradient approximation made simple. *Phys. Rev. Lett.* **1996**, *77*, 3865–3868. [CrossRef] [PubMed]

32. Miller, G.J.; Zhang, Y.; Wagner, F.R. Chemical bonding in solids. In *Handbook of Solid State Chemistry*; Dronskoski, R., Kikkawa, S., Stein, A., Eds.; Wiley-VCH: Weinheim, Germany, 2017; Volume 5, pp. 405–489.

33. Savin, A.; Nesper, R.; Wengert, S.; Fässler, T.F. ELF: The electron localization function. *Angew. Chem. Int. Ed.* **1997**, *36*, 1808–1832. [CrossRef]

34. Kokalj, A. Program XCRYSDEN. *J. Mol. Graph. Model.* **1999**, *17*, 176–178. [CrossRef]

35. Finger, L.W.; Kroeker, M.; Toby, B.H. DRAWXTL: An open-source computer program to produce crystal structure drawings. *J. Appl. Crystallogr.* **2007**, *40*, 188–192. [CrossRef]

36. Bader, R.W.F. *Atoms in Molecules. A Quantum Theory*; International Series of Monographs on Chemistry; Clarendon Press: Oxford, UK, 1994.

37. De-la Roza, A.O.; Blanco, M.A.; Martá, A.; Pendás, A.M.; Luaña, V. Program CRITIC2 (VERS. 1.0). *Comput. Phys. Commun.* **2009**, *180*, 157–166.

38. De-la Roza, A.O.; Luaña, V. A fast and accurate algorithm for QTAIM integrations in solids. *J. Comput. Chem.* **2010**, *32*, 291–305. [CrossRef] [PubMed]

39. FIZ Karlsruhe. *Inorganic Crystal Structure Database*; FIZ Karlsruhe: Karlsruhe, Germany, 2017.

40. Cirafici, S.; Franceschi, E. Stacking of close-packed AB$_3$ layers in RGa$_3$ compounds (R = heavy rare earth). *J. Less Common Met.* **1981**, *77*, 269–280. [CrossRef]

41. Larson, A.C.; Cromer, D.T.; Stambaugh, C.K. The crystal structure of PuAl$_3$. *Acta Crystallogr.* **1957**, *10*, 443–446. [CrossRef]

42. Shannon, R.D. Revised effective ionic radii and systematic studies of interatomic distances in halides and chalcogenides. *Acta Crystallogr.* **1976**, *A32*, 751–767. [CrossRef]

43. Pearson, W.B. *The Crystal Chemistry and Physics of Metals and Alloys*; Wiley Interscience: Hoboken, NJ, USA, 1972.

44. Wendorff, M.; Röhr, C. Alkaline-earth tri-mercurides A^{II}Hg$_3$ (A^{II}=Ca, Sr, Ba): Binary intermetallic compounds with a common and a new structure type. *Z. Kristallogr.* **2018**, in press.

45. Wade, K. Structural and bonding patterns in cluster chemistry. *Adv. Inorg. Chem. Radiochem.* **1976**, *18*, 1.

46. Jemmis, E.D.; Balakrishnarajan, M.M.; Pancharatna, P.D. Electronic requirements for macropolyhedral boranes. *Chem. Rev.* **2002**, *102*, 93–144. [CrossRef] [PubMed]

47. Todorov, E.; Sevov, S.C. Deltahedral clusters in neat solids: Synthesis and structure of the Zintl phase Cs$_4$Pb$_9$ with discrete Pb$_9^{4-}$ clusters. *Inorg. Chem.* **1998**, *37*, 3889–3891. [CrossRef] [PubMed]

48. Fässler, T.F. The renaissance of homoatomic nine-atom polyhedra of the heavier carbon-group element Si-Pb. *Coord. Chem. Rev.* **2001**, *215*, 347–377. [CrossRef]

49. Hoch, C.; Röhr, C.; Wendorff, M. Crystal structure of K$_4$Sn$_9$. *Acta Crystallogr.* **2002**, *C58*, 45–46.

50. Bobev, S.; Sevov, S.C. Isolated deltaheral clusters of leadin the solid state: synthesis and characterization of Rb$_4$Pb$_9$ and Cs$_{10}$K$_6$Pb$_{36}$ with Pb$_9^{4-}$, and A$_3'$A''Pb$_4$ (A' = Cs, Rb, K; A'' = Na, Li) with Pb$_4^{4-}$. *Polyhedron* **2002**, *21*, 641–649. [CrossRef]

51. Fässler, T.F.; Hunziker, M. The nido-Pb$_9^{4-}$ and the Jahn-Teller Distorted closo-Pb$_9^{3-}$ Zintl-Anions: Syntheses, X-ray Structures and Theoretical Studies. *Inorg. Chem.* **1995**, *33*, 5798–5809.

52. Hoch, C.; Wendorff, M.; Röhr, C. Synthesis and Crystal Structure of the Plumbides A$_4$Pb$_9$ and the Tetrelides A$_{12}$M$_{17}$ (A=Na, K, Rb, Cs; M=Si, Ge, Sn). *J. Alloys Compd.* **2003**, *361*, 206–221. [CrossRef]

crystals

MDPI

Article

Rhombohedral Distortion of the Cubic MgCu2-Type Structure in Ca2Pt3Ga and Ca2Pd3Ga

Asa Toombs [1,2] and Gordon J. Miller [1,2,*]

[1] Department of Chemistry, Iowa State University, Ames, IA 50011-3111, USA; toombsa@iastate.edu
[2] Ames Laboratory, U.S. Department of Energy, Ames, IA 50011-3111, USA
* Correspondence: gmiller@iastate.edu

Received: 27 March 2018; Accepted: 24 April 2018; Published: 26 April 2018

Abstract: Two new fully ordered ternary Laves phase compounds, Ca_2Pt_3Ga and Ca_2Pd_3Ga, have been synthesized and characterized by powder and single-crystal X-ray diffraction along with electronic structure calculations. Ca_2Pd_3Ga was synthesized as a pure phase whereas Ca_2Pt_3Ga was found as a diphasic product with Ca_2Pt_2Ga. Electronic structure calculations were performed to try and understand why $CaPt_2$ and $CaPd_2$, which crystalize in the cubic $MgCu_2$-type Laves phase structure, distort to the ordered rhombohedral variant, first observed in the magneto-restricted $TbFe_2$ compound, with the substitution of twenty-five percent of the Pt/Pd with Ga. Electronic stability was investigated by changing the valence electron count from $22e^-$/f.u. in $CaPd_2$ and $CaPt_2$ (2x) to $37e^-$/f.u. in Ca_2Pd_3Ga and Ca_2Pt_3Ga, which causes the Fermi level to shift to a more energetically favorable location in the DOS. The coloring problem was studied by placing a single Ga atom in each of four tetrahedra of the cubic unit cell of the $MgCu_2$-type structure, with nine symmetrically inequivalent models being investigated. Non-optimized and optimized total energy analyses of structural characteristics, along with electronic properties, will be discussed.

Keywords: coloring problem; band structure; structure optimizations; polar intermetallics; ternary Laves phases; electronic structure; X-ray diffraction; total energy

1. Introduction

Platinum exhibits the second largest relativistic contraction of the 6s orbitals, second only to Au [1], and has interesting catalytic properties, thus making it an interesting metal to study, especially in reduced environments. Exploration of the Ca-Pt-X (X = Ga, In) system revealed a new ternary compound, Ca_2Pt_3Ga, which turned out to be an ordered rhombohedral distortion of the $MgCu_2$-type Laves phase $CaPt_2$. After Ca_2Pt_3Ga was synthesized, the Pd analog was targeted and successfully prepared as an isostructural compound.

Laves phases are the largest class of intermetallic compounds with the general formula AB_2. Binary Laves phases primarily crystalize in one of three crystal structures: $MgCu_2$, $MgZn_2$, or $MgNi_2$ [2]. These structure types, which appear in a vast number of intermetallic systems, contain topologically closed-packed Frank–Kasper polyhedra and have been intensely studied for magnetic and magnetocaloric properties, hydrogen storage, and catalytic behavior [3–9]. The stability of Laves phases has been described by using geometrical and electronic structure factors. Geometrical factors include packing densities and atomic size ratios of A/B whereas electronic factors include valence electron count (*vec*) and difference in electronegativities between A and B [10,11].

To tune the properties and structures of Laves phases, an addition via substitution of a third element into the structure can be explored. The resulting ternary compounds can have mixing on the A and B sites but retain the parent Laves phase structure or the structures can change into a fully ordered ternary Laves phase that is a variant of the parent structure. The first ordered ternary Laves phase with the direct rhombohedral distortion of the $MgCu_2$-type, first evidenced in the binary $TbFe_2$ phase, which distorts via magnetostriction, was reported to be Mg_2Ni_3Si [11,12]. Recently, many examples of this rhombohedral distortion, with varying *vec*, have been reported: RE_2Rh_3Si (RE = Pr, Er), U_2Ru_3Si, $Ce_2Rh_{3+x}Si_{1-x}$, RE_2Rh_3Ge, Ca_2Pd_3Ge, Sm_2Rh_3Ge, U_2Ru_3Ge, Mg_2Ni_3Ge [2], Mg_2Ni_3P, and the heretofore only known gallides RE_2Rh_3Ga (RE = Y, La-Nd, Sm, Gd-Er).

The maximal *translationengleiche* subgroup of the Fd-3m space group that allows for the ordered 3:1 ratio of M:Ga atoms is R-3m [2,3,11]. Reported herein is the first extensive computational study of the "coloring problem" [13] for this rhombohedral distortion of the cubic cell. Formation energies and electronic structure calculations were performed on the binary and ternary Laves phases CaM_2 and Ca_2M_3Ga (M = Pt, Pd) along with total energy calculations of various "coloring models" with the same composition but different atomic arrangements or "colorings".

2. Materials and Methods

2.1. Electronic Structure Calculations

Electronic structure calculations to obtain DOS and COHP curves for bonding analysis were performed on CaM_2 and Ca_2M_3Ga (M = Pd, Pt) using the Stuttgart Tight-Binding, Linear-Muffin-Tin Orbital program with the Atomic Sphere Approximation (TB-LMTO-ASA) [14]. This approximation uses overlapping Wigner-Seitz (WS) spheres surrounding each atom so that spherical basis functions, i.e., atomic orbital (AO)-like wavefunctions, are used to fill space. The WS sphere radii used for the various atoms are: Ca 3.403–3.509 Å, Pt 2.800–2.844 Å, Pd 2.864–2.865 Å, and Ga 2.800–2.951 Å. The total overlap volume was 6.44% for the cubic Laves phase structures and 7.25% and 7.11%, respectively, for Ca_2Pt_3Ga and Ca_2Pd_3Ga. No empty spheres were required to attain 100% space filling of the unit cells. The exchange-correlation potential was treated with the von Barth-Hedin formulation within the local density approximation (LDA) [15]. All relativistic effects except spin-orbit coupling were taken into account using a scalar relativistic approximation [16]. The basis sets included 4s/(4p)/3d for Ca, 6s/6p/5d/(5f) for Pt, 5s/5p/4d/(4f) for Pd, and 4s/4p/(4d) for Ga (down-folded orbitals are shown in parentheses). Reciprocal space integrations were performed using *k*-point meshes of 1313 points for rhombohedral Ca_2Pt_3Ga and Ca_2Pd_3Ga and 145 points for cubic $CaPt_2$ and $CaPd_2$ in the corresponding irreducible wedges of the first Brillouin zones.

Structure optimization and total energy calculations were performed using the Vienna ab Initio Simulation Package (VASP) [17,18], which uses projector augmented-wave (PAW) [19] pseudopotentials that were treated with the Perdew–Burke–Ernzerhof generalized gradient approximation (PBE-GGA) [20]. Reciprocal space integrations were accomplished over a $13 \times 13 \times 13$ Monkhorst–Pack *k*-point mesh [21] by the linear tetrahedron method [22] with the energy cutoff for the plane wave calculations set at 500.00 eV.

2.2. Synthesis

All ternary compounds were obtained by melting mixtures of Ca pieces (99.99%, Sigma-Aldrich, St. Louis, MO, USA), Pt spheres (99.98%, Ames Laboratory, Ames, IA, USA) or Pd pieces (99.999%, Ames Laboratory), Ga ingots (99.99%, Alfa Aesar, Haverhill, MA, USA), and, in some cases, including Ag powder (99.9% Alfa Aesar). Samples of total weight ca. 300 mg were weighed out in a N_2-filled glovebox with <0.1 ppm moisture and sealed in tantalum tubes by arc-melting under an argon atmosphere. To prevent the tantalum tubes from oxidizing during the heating process, they were enclosed in evacuated silica jackets. The silica ampoules were placed in programmable furnaces and heated at a rate of 150 °C per hour to 1050 °C, held there for 3 h, then cooled at 50 °C per hour to 850 °C and annealed for 5 days. The samples were then cooled at 50 °C per hour to room temperature. Ca_2Pt_3Ga crystals were obtained from a loading of "$Ca_2Pt_3Ga_{0.85}Ag_{0.15}$", whereas Ca_2Pd_3Ga crystals were obtained from stoichiometric mixtures of these elements.

2.3. Powder X-ray Diffraction

All samples were characterized by powder X-ray diffraction on a STOE WinXPOW powder diffractometer using Cu $K\alpha$ radiation (λ = 1.540598 Å). Each sample was ground using an agate mortar and pestle and then sifted through a US standard sieve with hole sizes of 150 microns. All powder specimens were fixed on a transparent acetate film using a thin layer of vacuum grease, covered by a second acetate film, placed in a holder and mounted in the X-ray diffractometer. Scattered intensities were recorded using a scintillation detector with a step size of 0.03° in 2Θ using a step scan mode ranging from 0° to 130°. Phase analysis was performed using the program PowderCell [23] by overlaying theoretical powder patterns determined from single crystal X-ray diffraction over the powder patterns determined experimentally.

2.4. Single Crystal X-ray Diffraction

Single crystals were extracted from samples and mounted on the tips of thin glass fibers. Intensity data were collected at room temperature on a Bruker SMART APEX II diffractometer with a CCD area detector, distance set at 6.0 cm, using graphite monochromated Mo $K\alpha$ radiation (λ= 0.71073 Å). The data collection strategies were obtained both by a pre-saved set of runs as well as from an algorithm in the program COSMO in the APEX II software package [24]. Indexing and integration of data were performed with the program SAINT in the APEX II package [24,25]. SADABS was used to apply empirical absorption corrections [24]. Crystal structures were solved by direct methods and refined by full-matrix least squares on F^2 using SHELXL [26]. The final refinements were performed using anisotropic displacement parameters on all atoms. All crystal structure figures were produced using the program Diamond [27].

3. Results

3.1. Phase Analysis

Ca_2Pt_3Ga has been synthesized, but not as a pure phase. Using high temperature reactions, this compound can only be synthesized with the addition of a small quantity of silver in the loaded reaction mixture. Without adding silver the binary cubic Laves phase $CaPt_2$ forms as the major product as well as a small amount of some unidentified phase(s) according to X-ray powder diffraction. At reaction and annealing temperatures, Ga, Ca, and Ag are molten and their respective binary phase diagrams indicate miscibility. As Pt dissolves into the liquid, we hypothesize that the presence of Ag slows the reaction rate between Ca and Pt to form $CaPt_2$ and allows Pt and Ga to achieve sufficient mixing in the liquid before crystallizing into Ca_2Pt_3Ga. However, even when synthesized with the addition of silver and at 1050 °C, Ca_2Pt_3Ga does coexists with Ca_2Pt_2Ga and the binary $CaPt_2$. Figure 1 shows the powder pattern for the sample loaded as "$Ca_2Pt_3Ga_{0.85}Ag_{0.15}$" along with the theoretical patterns for the Ca_2Pt_3Ga and Ca_2Pt_2Ga. Peak fitting analysis performed with Jana [28] can be found in Figure S1 along with relative contributions of the three phases.

Three samples with varying silver content, viz., 2.50, 8.33, and 14.29 mole percent Ag, were examined. According to the XPD patterns (see Figure S3 in Supporting Information), as the silver content increased, the yield of Ca_2Pt_2Ga increased. The outcome was also successfully reproducible for the loading "$Ca_2Pt_3Ga_{0.85}Ag_{0.15}$" (see Figure S4).

On the other hand, Ca_2Pd_3Ga can be prepared without the addition of silver. Since Ag and Pd have similar X-ray scattering factors, using Ag for the synthesis of Ca_2Pd_3Ga would present distinct challenges for subsequent characterization. Fortunately, a stoichiometric loading yielded Ca_2Pd_3Ga essentially as a pure phase product, with two non-indexed peaks at 2θ values of 20.0° and 32.8° most likely coming from a slight impurity of $Ca_3Pd_2Ga_2$ (see Figure 2). Figure S2 shows the Rietveld refinement of Ca_2Pt_3Ga. Both compounds Ca_2Pt_3Ga and Ca_2Pd_3Ga degrade in air after approximately two weeks.

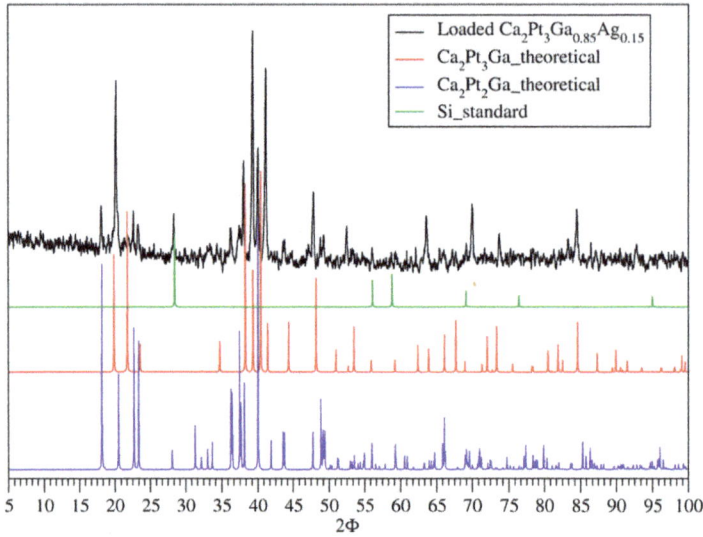

Figure 1. Powder Patterns for Ca_2Pt_3Ga showing the existence of Ca_2Pt_2Ga in the system.

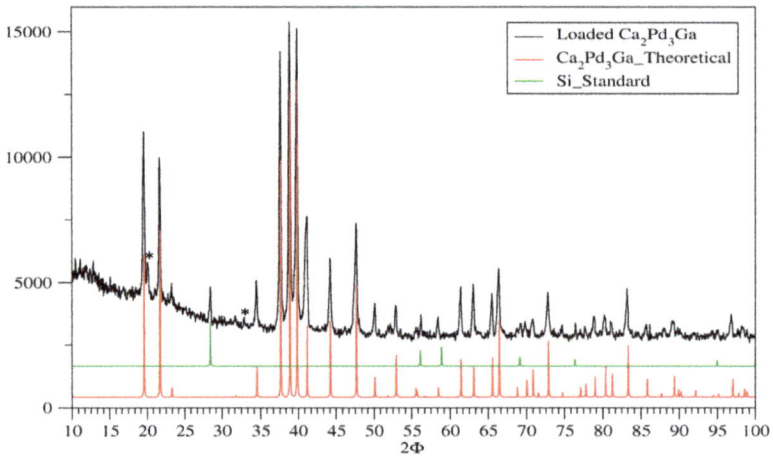

Figure 2. Powder pattern of sample loaded as Ca_2Pd_3Ga. * = non-indexed peaks coming from slight $Ca_3Pd_2Ga_2$ impurity.

3.2. Structure Determination

Several single crystals were selected from the loaded samples and tested on a Bruker SMART APEX II single crystal diffractometer for quality. Among those, the best 3 samples for Ca_2Pt_3Ga and Ca_2Pd_3Ga were examined. The results of single crystal X-ray diffraction on these crystals are found in Table 1 and Table S3. Ca_2M_3Ga (M = Pt, Pd) crystallize in the Y_2Rh_3Ge structure type [29], which is a rhombohedral distortion of the cubic Laves phase $MgCu_2$-type structure, as discussed by Doverbratt et al. [3] and Siggelkow et al. [2]. The M atoms occupy *9e* Wyckoff positions which buildup 3.6.3.6 Kagomé nets orthogonal to the *c*-axis [2]. The Ga atoms cap alternate triangular faces of the Kagomé net and themselves form hexagonal nets. The Ga and M atoms are each coordinated by 6 Ga or M atoms. Nine M and three Ga atoms surround each Ca atom in a slightly distorted Frank-Kasper polyhedral environment. The rhombohedral structure of Ca_2M_3Ga has no Ga-Ga nearest neighbor interactions, with distances ($6\times/6\times$) of 5.2361(9)/5.576(1) Å and 5.2331(6)/5.6326(8) Å for the Pt and Pd cases, respectively [3]. For the binary compounds CaM_2, each type of nearest neighbor interaction, i.e., Ca-Ca, Ca-M, and M-M, has a single length. When the structure distorts rhombohedrally to incorporate Ga atoms, the Ca-Ca and Ca-M contacts each split into two different distances: there are three shorter and one longer Ca-Ca interactions; whereas six Ca-M interactions that are longer and three shorter. Atomic positions and anisotropic displacement parameters for both compounds can be found in Tables S1 and S2.

Table 1. Crystallographic information for Ca_2Pt_3Ga and Ca_2Pd_3Ga.

Sample	Ca_2Pt_3Ga	Ca_2Pd_3Ga
Space Group	R-3m	R-3m
Unit Cell Dim.	$a = 5.576(1)$ Å $c = 12.388(3)$ Å	$a = 5.6326(8)$ Å $c = 12.300(2)$ Å
Volume	333.6(2) Å^3	337.9(1) Å^3
Z	3	3
Theta range for data collection	4.530 to 28.922°	4.495 to 49.319°
Index ranges	$-7 \leq h \leq 7, -7 \leq k \leq 7, -16 \leq l \leq 16$	$-11 \leq h \leq 11, -11 \leq k \leq 11, -26 \leq l \leq 26$
Reflections Collected	1321	4878
Independent Reflections	126 [R(int) = 0.0485]	460 [R(int) = 0.0585]
Data/restraints/parameters	126/0/11	460/0/11
Goodness-of-fit	1.078	1.091
Final R indices [I > 2sigma(l)]	R1 = 0.0179, wR2 = 0.0403	R1 = 0.0208, wR2 = 0.0403
R indices (all data)	R1 = 0.0179, wR2 = 0.0403	R1 = 0.0251, wR2 = 0.0414
Extinction Coefficient	0.0008(1)	0.0036(3)
Largest diff. peak and hole	3.366 and -2.052 e·Å^{-3}	1.697 and -2.657 e·Å^{-3}

3.3. Coloring Models and Electronic Structure Calculations

To elucidate the driving force that leads to the rhombohedral arrangement of Pt/Pd and Ga atoms in the majority atom positions of the cubic Laves phase, Burnside's lemma [30] was used to construct nine distinct structural models for the unit cells formulated as $Ca_8[M_3Ga]_4$ [31]. These models are illustrated in Figure S5 along with their respective space groups. The total energy of each model, listed in Table 2, was calculated using VASP without optimization. These results show that the rhombohedral model (α) is the lowest in energy. There are three models that have no nearest neighbor Ga-Ga interactions, these include the rhombohedral coloring, a cubic coloring (μ), and an orthorhombic coloring (γ). The rhombohedral and cubic models are the two lowest in energy before optimization, whereas the orthorhombic coloring is the second highest in energy.

Table 2. Total energy of non-optimized and optimized coloring models showing optimized Ga-Ga distances.

Model	Ca_2Pt_3Ga				Ca_2Pd_3Ga		
	Ga-Ga Distance (Å)	meV Non-Optimized	meV Optimized	# Ga-Ga Interactions	meV Optimized	meV Non-Optimized	Ga-Ga Distance (Å)
Alpha	5.276	0	0	0	0	0	5.282
Beta	2.755	+38.5	+78.8	1	+68.1	+30.5	2.732
Delta	2.811	+62.1	+121.8	2	+121.8	+59.3	2.775
Epsilon	2.808	+50.8	+95.7	1	+73.0	+36.0	2.800
Gamma	4.738	+140.7	+56.0	0	+26.0	+92.0	4.791
Iota	2.931	+70.5	+113.4	2	+104.7	+61.2	2.922
Mu	4.709	+37.6	+71.0	0	+66.5	+13.3	4.653
Theta	2.950	+203.0	+289.4	6	+256.6	+155.5	2.909
Zeta	2.824	+97.6	+168.6	3	+150.0	+81.1	2.798

All models were then optimized using VASP by allowing all structural parameters, including volume, cell shape, and atomic coordinates, to relax. The energies of the optimized structures can also be found in Table 2. Details of the optimized structures can be found in Tables S5–S13. After optimization, the rhombohedral model remains the lowest energy with the orthorhombic and cubic models coming in second and third lowest energy. The orthorhombic (γ) model optimized to a monoclinic unit cell, $a \neq b \neq c$ and only $\beta = 90°$, while the other two models retained their symmetry through the optimization. As the energies increase for the various optimized structures, the number of nearest neighbor Ga-Ga interactions increases and the corresponding Ga-Ga distances decrease. Table 2 lists the energies after optimization as well as the resulting nearest Ga-Ga distance showing the correlation between total energy and Ga-Ga interactions.

Calculations have shown that the rhombohedral model is the lowest energy both before and after optimization. This leads to the question of how does the rhombohedral structure arise from the "cubic" arranged starting point? The "cubic" starting point, for Ca_2Pt_3Ga, has a lattice parameter of 7.598 Å and a body diagonal distance of 13.160 Å. The primitive rhombohedral unit cell, within the face-centered cubic cell, has angles equal to 60° before optimization. After optimization these angles are 57.825° and 57.515° for Ca_2Pt_3Ga and Ca_2Pd_3Ga respectively. This collapsing of the angles, comprising the primitive unit cell, causes the expansion of the body diagonal. Optimization expands the lattice parameter by 1.43% to 7.708 Å. However, during optimization, the body diagonals do not expand congruently. Three body diagonals are congruent with a length of 13.627 Å while the fourth body diagonal shrinks to 12.482 Å. In the Pd case the three equivalent body diagonals are 13.754 Å while the fourth body diagonal shrinks to 12.412 Å. The shorter body diagonal is parallel to the direction that the Ga atoms are located in each of the four tetrahedra in the unit cell (see Figure 3). Calculated ICOHP values for the "cubic" starting point indicate that the Pt-Ga bonds (ICOHP = −1.857 eV) would become shorter as compared with the Pt-Pt bonds (ICOHP = −0.775 eV). These ICOHP values shed light on the driving force behind the corresponding distortion of the cubic unit cell to the rhombohedral structure by substituting Ga for Pd/Pt: shorter Pt-Ga contacts will contract one body-diagonal of the cubic cell.

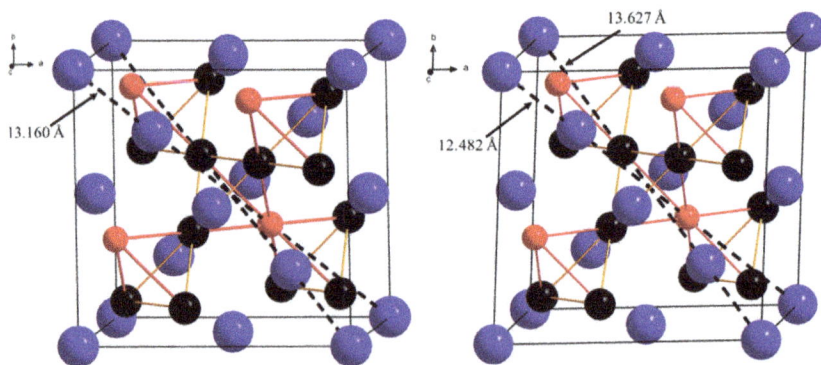

Figure 3. "Cubic" arranged starting point with 4 congruent body diagonals (**left**) optimized rhombohedral unit cell (**right**) with 3 body diagonals of 13.627 Å and 1 body diagonal of 12.482 Å. Red colored bonds shrink during optimization causing the reduction in body diagonal length from 011 to 100. Blue, black, and red atoms are Ca, Pt, and Ga respectively.

All "coloring" calculations performed above were evaluated for the case of $vec = 37e^-$/f.u. Models were also created for the Ca_2Pd_3Ge to determine if there were any differences by changing the vec to $38\ e^-$/f.u. The energies of the non-optimized and optimized colorings can be found in Table S4, and the results show the same trends as those for Ca_2M_3Ga (M = Pd, Pt): the lowest energy model shows rhombohedral coloring and the largest Ge-Ge nearest neighbor distance.

To shed light on the nature of the distortion of the cubic Laves phase structure of CaM_2 (M = Pd, Pt) upon the substitution of Ga for M and their stability, various total energies as well as the density of states (DOS) and crystal orbital Hamilton population (COHP) curves were determined for the binaries CaM_2 and ternaries Ca_2M_3Ga (M = Pd,Pt). With respect to the elemental solids, the binaries CaM_2 and ternaries Ca_2M_3Ga are favored. For the experimental structures, $CaPt_2$ has an especially high formation energy compared to $CaPd_2$; the experimental volume of $CaPt_2$ is ca. 10 $Å^3$/formula unit smaller than that of $CaPd_2$, although Pt is slightly larger than Pd (the 12-coordinate metallic radii are 1.39 (Pt) and 1.37 (Pd) [32]. This outcome may be an indication of relativistic effects on the valence orbitals of Pt, effects which can enhance empty d-filled d Ca-Pt $3d$-$5d$ interactions as compared to Ca-Pd $3d$-$4d$ interactions.

Total energy calculations for the "reactions" $2CaPt_2 + Ga \rightarrow Ca_2Pt_3Ga + Pt$ and $2CaPd_2 + Ga \rightarrow Ca_2Pd_3Ga + Pd$ illuminate significant aspects of the relative stability and formation of the ternaries. Both "reactions" yield favorable ΔE values, respectively, of -0.297 and -0.599 eV, and they can be separated into the sum of three hypothetical but revealing steps: (1) elemental substitution of Ga for M with no metric changes (volume, unit cell shape) in the cubic Laves phase structure; (2) volume change for the ternary without change in unit cell shape; and (3) distortion of the unit cell shape of the ternary at constant cell volume. Table 3 summarizes these results. The difference in overall total energies essentially arises from the energetic difference for step (1) between Pt and Pd cases: replacing Pt with Ga is energetically unfavorable whereas Pd for Ga is favorable. Since Ga (metallic radius of 1.50 Å [32]) is larger than either Pt or Pd, these energy differences reflect, in part, volume effects, but also are influenced by the relative Mulliken electronegativities [33], which increase Ga (3.2 eV) to Pd (4.45 eV) to Pt (5.6 eV). In the second step, both structures favorably expand by over 3% (based on unit cell volume) with slight energetic lowering. In the third step, both lattice distortions provide similar energetic stabilizations.

Table 3. Relative total energies (eV) for conversion of binary cubic Laves phases CaM_2 into ternary rhombohedral derivatives Ca_2M_3Ga (M = Pd, Pt). * = cubic unit cell metric with volume matching the corresponding binary compound; ** = cubic unit cell metric with volume matching the corresponding rhombohedral ternary compound. Specific total energies (eV/formula unit) for each component are listed in Supporting Information.

	$2CaPd_2$	+ Ga	$\rightarrow$	Ca_2Pd_3Ga	+ Pd	$\Delta E = -0.599$ eV
(1)	$2CaPd_2$	+ Ga	$\rightarrow$	Ca_2Pd_3Ga *	+ Pd	$\Delta E = -0.258$ eV
(2)	Ca_2Pd_3Ga *		$\rightarrow$	Ca_2Pd_3Ga **		$\Delta E = -0.042$ eV
(3)	Ca_2Pd_3Ga **		$\rightarrow$	Ca_2Pd_3Ga		$\Delta E = -0.299$ eV
	$2CaPt_2$	+ Ga	$\rightarrow$	Ca_2Pt_3Ga	+ Pt	$\Delta E = -0.297$ eV
(1)	$2CaPt_2$	+ Ga	$\rightarrow$	Ca_2Pt_3Ga *	+ Pt	$\Delta E = +0.115$ eV
(2)	Ca_2Pt_3Ga *		$\rightarrow$	Ca_2Pt_3Ga **		$\Delta E = -0.088$ eV
(3)	Ca_2Pt_3Ga **		$\rightarrow$	Ca_2Pt_3Ga		$\Delta E = -0.324$ eV

We now consider some specific aspects of the electronic structures of the binary CaM_2 and ternary Ca_2M_3Ga. The DOS and COHP for $CaPt_2$ are shown in Figure 4; the DOS shows a significant sharp peak at the Fermi level which suggests a potential electronic instability.

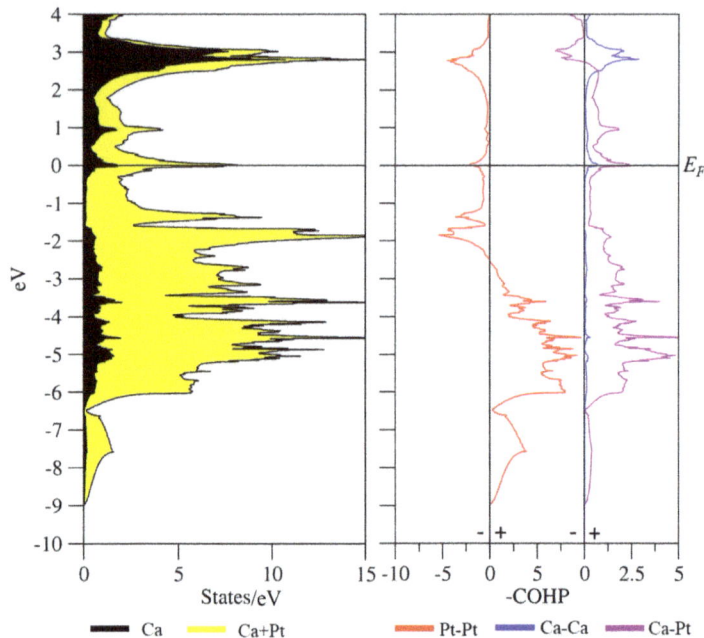

Figure 4. DOS (**left**) and -COHP (**right**) for CaPt$_2$ with +/− indicating bonding/antibonding values of COHP curves. Ca-Ca and Ca-Pt interactions are magnified for comparison.

The precise reason for this peak and the possible instability has not been forthcoming to date and an electronic structure investigation was performed. Analysis of the band structure at the Fermi level, specifically at the W point, because there are bands with nearly zero slope near this point, shows that the Pt 5d atomic orbitals are major contributors to this peak. For each Pt atom the orbital contribution is split approximately as 80 percent 5d and 20 percent 6p, and the primary interatomic orbital interactions are 5d-5d π^* interactions. Spin-orbit coupling was added to the Hamiltonian operator of CaPt$_2$ to see if the peak at the Fermi level would be affected. Figure S6 shows the DOS curve and band structure for these calculations. The degeneracies in the band structure at the high symmetry k-points are removed by the spin-orbit coupling; however, the peak at E_F is not affected. Furthermore, the application of spin-orbit coupling would only converge with a unit cell expanded by ~1.1 Å. A second hypothesis was to consider that vacancies might be playing a role to stabilize the structure. DOS calculations were subsequently performed using VASP on "Ca$_8$Pt$_{15}$" and "Ca$_8$Pt$_{12}$" models to see the effect of vacancies on the peak at E_F; these results can be seen in Figure S7. The peak is no longer present in the "Ca$_8$Pt$_{12}$" model only when all 4 vacancies are adjacent to each other so that the resulting defect Laves phase structure is missing an entire tetrahedron per unit cell. Lastly, DOS curves calculated for AEM$_2$ (AE = Ca, Sr, Ba; M = Ni, Pd, Pt), shown in Figure S8, indicate that as the size of the alkaline earth metal increases, the peak in the DOS moves away from E_F. These calculations have not identified the reason behind the potential electronic instability in CaM$_2$ and now the possibility of superconductivity is being investigated.

The Pt 5d band dominates the overall occupied DOS of CaPt$_2$ and the majority contribution to the overall polar covalent bonding via ICOHP values arises from Pt-Pt interactions, which switch from bonding to antibonding at an energy value corresponding to 12.42 e^-/formula unit. At the Fermi level, the Pt-Pt COHP curve shows a small antibonding peak, whereas the Ca-Pt and Ca-Ca COHPs show a small bonding peak. On transitioning to Ca$_2$Pt$_3$Ga, the peak at the Fermi level in the DOS of CaPt$_2$ has all but disappeared in the DOS curve of Ca$_2$Pt$_3$Ga (see Figure 5). There is a large range from ca.

−1.5 to +1 eV over which the DOS of Ca_2Pt_3Ga remains flat. The COHP curves have also changed in that the Pt-Pt interactions still transition to antibonding at −2.60 eV, although around the Fermi level they are mostly nonbonding rather than antibonding. Overall, at the Fermi level there are now no small peaks but mostly constant intensities of nonbonding or slightly bonding interactions with the Ca-Pt interactions still contributing the most to overall bonding.

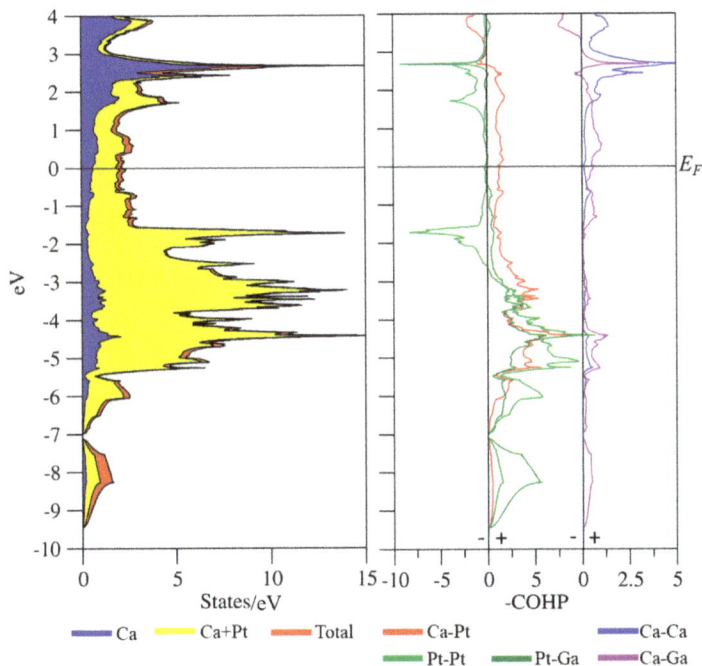

Figure 5. DOS (**left**) and COHP (**right**) for Ca_2Pt_3Ga with +/− indicating bonding/antibonding values of COHP curves. Ca-Ca and Ca-Ga interactions magnified for comparison.

The DOS and COHP curves for $CaPd_2$ (see Figure 6) and Ca_2Pd_3Ga (see Figure 7) look similar overall to those described above for the Pt counterparts, although there are a few distinct differences. The peak at the Fermi level in the DOS for $CaPd_2$ as well as the antibonding Pd-Pd peak in the COHP are not as intense nor as sharp as those observed for $CaPt_2$. The Pd-Pd interactions transition from bonding to antibonding at an energy which corresponds to 15.36 eV/formula unit. There is a similar minimum in the Pd 4*d* states at −1.40 eV which corresponds to 7.5 4*d* electrons/Pd. The DOS and COHP for Ca_2Pd_3Ga are similar to those for Ca_2Pt_3Ga. There is again a large energy range from ca. −1.8–1.8 eV over which the DOS remains flat.

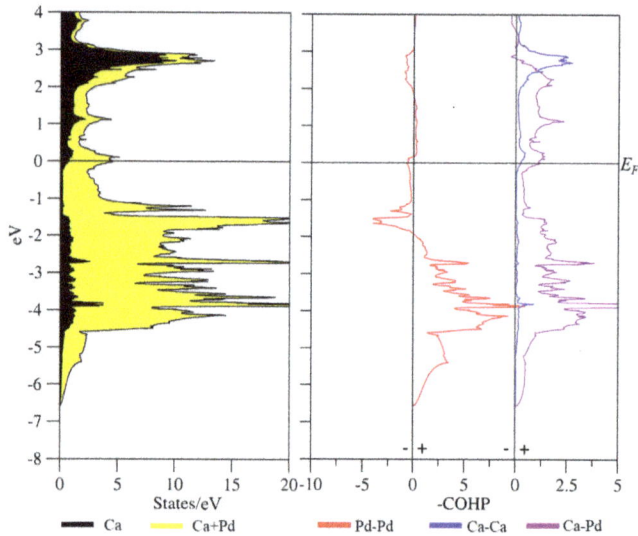

Figure 6. DOS (**left**) and COHP (**right**) calculated for $CaPd_2$ with $+/-$ indicating bonding/antibonding values of COHP curves. Ca-Pd and Ca-Ca interactions magnified for comparison.

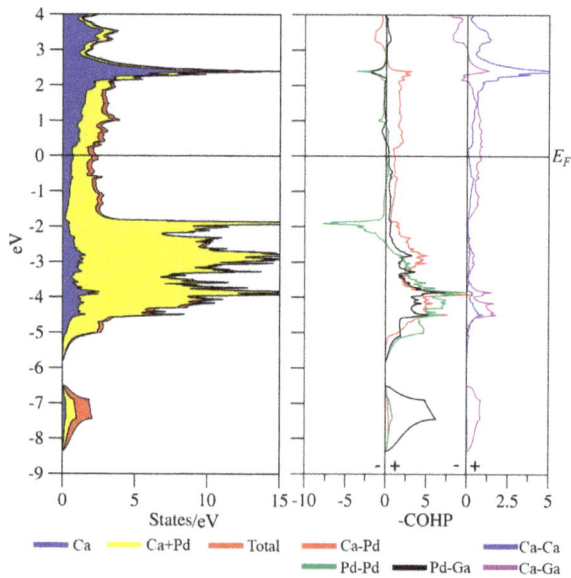

Figure 7. DOS (**left**) and COHP (**right**) calculated for Ca_2Pd_3Ga with $+/-$ indicating bonding/antibonding values of COHP curves. Ca-Ca and Ca-Ga interactions magnified for comparison.

Tables S14–S17 contain the percent contributions of each type of interaction to the overall integrated COHP (ICOHP) values. These can be used as an indication to determine the bond energy contribution to the stability of a structure. The cubic binary Laves phases are dominated by the M-M (M = Pd, Pt) interactions at 62.95 and 58.41% for $CaPt_2$ and $CaPd_2$, respectively, whereas the Ca-Ca interactions lend little to the overall ICOHP at only 1.47% and 2.12% for $CaPt_2$ and $CaPd_2$, respectively. The rest is made up by the Ca-M interactions. For the ternary compounds, there is a significant drop in the contributions from M-M interactions to the total polar-covalency because the majority of the total ICOHP comes from Ca-M and M-Ga interactions. In Ca_2Pt_3Ga, the Pt-Pt contribution is just 29.95%, and in Ca_2Pd_3Ga, the Pd-Pd contribution is 24.40%. The Ca-M and M-Ga interactions contribute ~30% each in both ternary compounds. Furthermore, in the ternary compounds, there are now Ca-Ga interactions that account for 5.81% and 7.89%, respectively, in Ca_2Pt_3Ga and Ca_2Pd_3Ga. The Ca-Ca bond populations continue to contribute less than 2.5% in all cases.

The analysis of the integrated Hamilton population curves to study the interatomic interactions is useful for mainly two-center interactions, and the use of crystal orbital overlap population (COOP) analysis has been applied to examine cubic vs. hexagonal Laves phases [34]. However, multi-center chemical bonding in Laves phases has also been described using an electron localizability indicator [35]. Laves phases AB_2 with a small electronegativity difference between the two components exhibit enhanced multi-center bonding because the effective charge transfer between A and B is small. However, if the electronegativity difference between A and B is large, the correspondingly greater effective charge transfer can lead to polyanions [35]. $CaPt_2$ and $CaPd_2$ are cases with a large electronegativity difference between the A and B atoms (3.2 eV for $CaPt_2$ and 2.25 eV for $CaPd_2$ [33]), but there most likely exists some multi-center bonding in these compounds, especially between the faces of the transition metal tetrahedra and the Ca atoms, bonding which would be affected by the substitution of one of the Pt or Pd atoms by Ga. The electronegativity difference between Ga and Ca is not as large (1.0 eV [33]), which creates less charge transfer between Ca and Ga and thereby shifts the valence electron density as compared to the binary structures. Electron localizability indicators were not evaluated during this study, but they would provide useful information regarding specifics of multi-center interactions in the ternary derivatives.

4. Conclusions

Two new fully ordered ternary gallium-containing Laves phase compounds, Ca_2Pt_3Ga and Ca_2Pd_3Ga, were synthesized and characterized by X-ray diffraction and electronic structure calculations. The compounds crystallize in a rhombohedrally distorted cubic $MgCu_2$-type structure. Ca_2Pd_3Ga was synthesized as a pure phase whereas Ca_2Pt_3Ga was only found co-existing with Ca_2Pt_2Ga and $CaPt_2$ and required the addition of small amounts of Ag to form. Total energy calculations performed on nine symmetrically inequivalent "coloring models" indicated that the rhombohedral coloring gave the lowest energy by 78.8 meV and 68.1 meV in the Pt and Pd cases respectively. ICOHP values from the "cubic" starting arrangement indicate that the M-Ga bonds tend to become shorter vs. M-M contacts. These shorter bonds cause a shrinking of the body diagonal from the corner of 011 to 100 which gives rise to the rhombohedrally distorted unit cell.

This computational study supports the group–subgroup relationship that R-3m is the highest maximal subgroup of Fd-3m and allows a fully ordered structure with the 3:1 ratio of M:Ga atoms. The rhombohedral coloring is the lowest energy way to color the "B" network in the Laves phase so that the Ga atoms avoid all homoatomic interactions and have the furthest nearest neighbor distance.

Supplementary Materials: The following are available online at http://www.mdpi.com/2073-4352/8/5/186/search.html. Ca_2Pt_3Ga: 3000193, Ca_2Pd_3Ga: 3000192.

Author Contributions: A.T. performed the synthesis, characterization, and computational studies of all compounds, and wrote the first draft of the manuscript as part of his doctoral dissertation. G.J.M. managed the overall project and edited the final manuscript.

Acknowledgments: A. Toombs acknowledges J. Pham and L. Lutz-Kappelman for answering many questions involving crystallography and computational guidance along with S. Thimmaiah for helping with the Rietveld refinements. This research was supported by the Office of the Basic Energy Sciences, Materials Sciences Division, U.S. Department of Energy (DOE). The Ames Laboratory is operated by Iowa State University under Contract No. DE-AC02-07C. Computations were performed on the CRUNCH cluster supported by Iowa State University Computation Advisory Committee under Project No. 202-17-10-08-0005.

Conflicts of Interest: The authors declare no conflict of interest.

References

1. Jansen, M. Effects of relativistic motion of electrons on the chemistry of gold and platinum. *Solid State Sci.* **2005**, *7*, 1464–1474. [CrossRef]
2. Siggelkow, L.; Hlukhyy, V.; Faessler, T.F. The Influence of the Valence Electron Concentration on the Structural Variation of the Laves Phases MgNi$_{2-x}$Ge$_x$. *Z. Anorg. Allg. Chem.* **2017**, *643*, 1424–1430. [CrossRef]
3. Doverbratt, I.; Ponou, S.; Lidin, S. Ca$_2$Pd$_3$Ge, a new fully ordered ternary Laves phases structure. *J. Solid State Chem.* **2013**, *197*, 312–316. [CrossRef]
4. Stein, F.; Palm, M.; Sauthoff, G. Structure and stability of Laves phases. Part I—Critcal assessment of factors controlling Laves phase stability. *Intermetallics* **2004**, *12*, 713–720. [CrossRef]
5. Osters, O.; Nilges, T.; Schöneich, M.; Schmidt, P.; Rothballer, J.; Pielnhofer, F.; Weihrich, R. Cd$_4$Cu$_7$As, The First Representative of a Fully Ordered, Orthorhombically Distorted MgCu$_2$ Laves Phases. *Inorg. Chem.* **2012**, *51*, 8119–8127. [CrossRef] [PubMed]
6. Murtaza, A.; Yang, S.; Zhou, C.; Song, X. Influence of Tb on easy magnetization direction and magnetostriction of ferromagnetic Laves phase GdFe compounds. *Chin. Phys. B* **2016**, *25*, 096107. [CrossRef]
7. Manickam, K.; Grant, D.M.; Walker, G.S. Optimization of AB2 type alloy composition with superior hydrogen storage properties for stationary applications. *Int. J. Hydrogen Energy* **2015**, *40*, 16288–16293. [CrossRef]
8. Nash, C.P.; Boyden, F.M.; Whittig, L.D. Intermetallic compounds of alkali metals with platinum. A novel preparation of a colloidal platinum hydrogenation catalyst. *J. Am. Chem. Soc.* **1960**, *82*, 6203–6204. [CrossRef]
9. Xiao, C.; Wang, L.; Maligal-Ganesh, R.V.; Smetana, V.; Walen, H.; Thiel, P.A.; Miller, G.J.; Johnson, D.D.; Huang, W. Intermetallic NaAu$_2$ as a Heterogeneous Catalyst for Low-Temperature CO Oxidation. *J. Am. Chem. Soc.* **2013**, *135*, 9592–9595. [CrossRef] [PubMed]
10. Thimmaiah, S.; Miller, G.J. Influence of Valence Electron Concentration on Laves Phases: Structures and Phase Stability of Pseudo-Binary MgZn$_{2-x}$Pd$_x$. *Z. Anorg. Allg. Chem.* **2015**, *641*, 1486–1494. [CrossRef]
11. Seidel, S.; Janka, O.; Benndorf, C.; Mausolf, B.; Haarmann, F.; Eckert, H.; Heletta, L.; Pöttgen, R. Ternary rhombohedral Laves phases RE$_2$Rh$_3$Ga (RE = Y, La-Nd, Sm, Gd-Er). *Z. Naturforsch.* **2017**, *72*, 289–303. [CrossRef]
12. Dwight, A.E.; Kimball, C.W. A rhombohedral Laves phase. Terbium-ron (TbFe$_2$). *Acta Cryst. Sect. B* **1974**, *30*, 2791–2793. [CrossRef]
13. Burdett, J.K.; Canadell, E.; Hughbanks, T. Symmetry control of the coloring problem: The electronic structure of MB2C2 (M = calcium, lanthanum, . . .). *J. Am. Chem. Soc.* **1986**, *108*, 3971–3976. [CrossRef]
14. Jepsen, O.; Andersen, O.K. *TB-LMTO*, version 47; Max-Planck-Institut für Festkörperforshcung: Stuttgart, Germany, 2000.
15. Von Barth, U.; Hedin, L. Local exchange-correlation potential for the spin-polarized case. *J. Phys. C* **1972**, *5*, 1629. [CrossRef]
16. Koelling, D.D.; Harmon, B.N. A technique for relative spin-polarized calculations. *J. Phys. C* **1977**, *10*, 3107–3114. [CrossRef]
17. Kresse, G.; Hafner, J. Ab initio molecular dynamics for liquid metals. *Phys. Rev. B* **1993**, *47*, 558. [CrossRef]
18. Kreese, G.; Furthmüller, J. Efficient iterative schemes for ab initio total-energy calculations using a plane-wave basis set. *Phys. Rev. B* **1996**, *54*, 11169–11186. [CrossRef]
19. Blöchl, P.E. Projector augmented-wave method. *Phys. Rev. B* **1994**, *50*, 17953–17979. [CrossRef]
20. Perdew, J.P.; Burke, K.; Wang, Y. Gerneralized gradient approximation for the exchange-correlation hole of a many-electron system. *Phys. Rev. B* **1996**, *54*, 16533–16539. [CrossRef]
21. Monkhorst, H.J.; Pack, J.D. Special points for Brillouin-zone integrations. *Phys. Rev. B* **1976**, *13*, 5188. [CrossRef]
22. Blöchl, P.E.; Jepsen, O.; Andersen, O.K. Improved tetrahedron method for Brillouin-zone integrations. *Phys. Rev. B* **1994**, *49*, 16223–16233. [CrossRef]
23. *PowderCell*, version 2.3; Federal Institute for Materials Research and Testing: Berlin, Germany, 2000.
24. *APEX-II*; Bruker AXS Inc.: Madison, WI, USA, 2013.
25. *SAINT-V8.27*; Bruker AXS Inc.: Madison, WI, USA, 2013.
26. *SHELXTL-v2008/4*; Bruker AXS Inc.: Madison, WI, USA, 2013.

27. Brandenburg, K. *Diamond*, version 3.2; Crystal Impact GbR: Bonn, Germany, 2011.

28. Petricek, V.; Dusek, M.; Palatinus, L. *Structure Determination Software Programs*; Institute of Physics: Praha, Czech Republic, 2006.

29. Cenzual, K.; Chabot, B.; Parthe, E. Yttrium rhodium germanide (Y_2Rh_3Ge), a rhombohedral substitution variant of the $MgCu_2$ type. *J. Solid State Chem.* **1987**, *70*, 229–234. [CrossRef]

30. Burdett, J.K.; McLarnan, T.J. Geometrical and electronic links among the structures of MX2 solids: Structural enumeration and electronic stability of pyritelike systems. *Inorg. Chem.* **1982**, *21*, 1119–1128. [CrossRef]

31. Xie, W.; Miller, G.J. β-Mn-Type $Co_{8+x}Zn_{12-x}$ as a Defect Cubic Laves Phase: Site Preferences, Magnetism, and Electronic Structure. *Inorg. Chem.* **2013**, *52*, 9399–9408. [CrossRef] [PubMed]

32. Wells, A.F. *Structural Inorganic Chemistry*, 5th ed.; Clarendon Press: Oxford, UK, 1984.

33. Pearson, R.G. Absolute Electronegativity and Hardness: Application to Inorganic Chemistry. *Inorg. Chem.* **1988**, *27*, 734–740. [CrossRef]

34. Johnston, R.; Hoffmann, R. Structure-Bonding Relationships in the Laves Phases. *Z. Anorg. Allg. Chem.* **1992**, *616*, 105–120. [CrossRef]

35. Ormeci, A.; Simon, A.; Grin, Y. Structural Topology and Chemical Bonding in Laves Phases. *Angew. Chem.* **2010**, *49*, 8997–9001. [CrossRef] [PubMed]

crystals

MDPI

Article

Crystal Structure, Spectroscopic Investigations, and Physical Properties of the Ternary Intermetallic $REPt_2Al_3$ (RE = Y, Dy–Tm) and $RE_2Pt_3Al_4$ Representatives (RE = Tm, Lu)

Fabian Eustermann [1], Simon Gausebeck [1], Carsten Dosche [2], Mareike Haensch [2], Gunther Wittstock [2] and Oliver Janka [1,2,*]

[1] Institut für Anorganische und Analytische Chemie, Universität Münster, Corrensstrasse 30, 48149 Münster, Germany; f_eust01@wwu.de (F.E.); simongausebeck@gmx.de (S.G.)
[2] Institut für Chemie, Carl von Ossietzky Universität Oldenburg, 26111 Oldenburg, Germany; carsten.dosche@uni-oldenburg.de (C.D.); mareike.haensch@uni-oldenburg.de (M.H.); gunther.wittstock@uni-oldenburg.de (G.W.)
* Correspondence: ocjanka@uni-muenster.de; Tel.: +49-251-83-36074

Received: 1 March 2018; Accepted: 10 April 2018; Published: 16 April 2018

Abstract: The $REPt_2Al_3$ compounds of the late rare-earth metals (RE = Y, Dy–Tm) were found to crystallize isostructural. Single-crystal X-ray investigations of YPt_2Al_3 revealed an orthorhombic unit cell (a = 1080.73(6), b = 1871.96(9), c = 413.04(2) pm, $wR2$ = 0.0780, 942 F^2 values, 46 variables) with space group $Cmmm$ ($oC48$; q^2pji^2hedb). A comparison with the Pearson database indicated that YPt_2Al_3 forms a new structure type, in which the Pt and Al atoms form a $[Pt_2Al_3]^{\delta-}$ polyanion and the Y atoms reside in the cavities within the framework. Via a group-subgroup scheme, the relationship between the $PrNi_2Al_3$-type structure and the new YPt_2Al_3-type structure was illustrated. The compounds with RE = Dy–Tm were characterized by powder X-ray diffraction experiments. While YPt_2Al_3 is a *Pauli*-paramagnet, the other $REPt_2Al_3$ (RE = Dy–Tm) compounds exhibit paramagnetic behavior, which is in line with the rare-earth atoms being in the trivalent oxidation state. $DyPt_2Al_3$ and $TmPt_2Al_3$ exhibit ferromagnetic ordering at T_C = 10.8(1) and 4.7(1) K and $HoPt_2Al_3$ antiferromagnetic ordering at T_N = 5.5(1) K, respectively. Attempts to synthesize the isostructural lutetium compound resulted in the formation of $Lu_2Pt_3Al_4$ ($Ce_2Ir_3Sb_4$-type, *Pnma*, a = 1343.4(2), b = 416.41(8), c = 1141.1(2) pm), which could also be realized with thulium. The structure was refined from single-crystal data ($wR2$ = 0.0940, 1605 F^2 values, 56 variables). Again, a polyanion with bonding Pt–Al interactions was found, and the two distinct Lu atoms were residing in the cavities of the $[Pt_3Al_4]^{\delta-}$ framework. X-ray photoelectron spectroscopy (XPS) measurements were conducted to examine the electron transfer from the rare-earth atoms onto the polyanionic framework.

Keywords: intermetallics; crystal structure; group-subgroup; magnetic properties; XPS

1. Introduction

In the field of intermetallic compounds [1,2], some structure types are found with an impressive number of entries listed in the Pearson database [3]. Amongst them are the binary Laves phases of the $MgCu_2$-type ($Fd\bar{3}m$) [4] and $MgZn_2$-type ($P6_3/mmc$) [5] structures (together with more than 5500 entries), the cubic Cu_3Au-type ($Pm\bar{3}m$, >1950 entries) structures [6], and the hexagonal $CaCu_5$-type ($P6/mmm$, >1650 entries) structures [7]. For ternary intermetallic compounds, the tetragonal body-centered $ThCr_2Si_2$-type ($I4/mmm$, >3250 entries) [8], the orthorhombic TiNiSi-type ($Pnma$, >1550 entries), and the hexagonal ZrNiAl-type ($P\bar{6}2m$, >1450 entries) [9] representatives show a broad variety of compounds with numerous, different elemental combinations. The structures and physical properties of the equiatomic *RETX* (*RE* = rare-earth element, *T* = transition metal, *X* = element of group 12–15) representatives have been recently summarized in a series of review articles [10–13].

Derived from the binary $CaCu_5$-type structure, two prototypic ternary representatives with different chemical compositions have been reported: the $CeCo_3B_2$- [14] and the $PrNi_2Al_3$-type [15] structures. From a crystal chemical point of view, YNi_2Al_3 is also worth mentioning [16], because this compound can be considered to be an i3-superstructure of the $PrNi_2Al_3$-type structure. Recently, an i7-superstructure of $PrNi_2Al_3$ has also been reported, which was also found for $ErPd_2Al_3$ [17]. Our interests in the compounds of the $REPt_2Al_3$ series originate from the fact that only $CePt_2Al_3$ ($PrNi_2Al_3$-type) has been reported previously [18]. Therefore, we synthesized and characterized the missing members of the $REPt_2Al_3$ series with the late, small rare-earth elements. From a basic research point of view, investigations of the magnetic ground state of the open *f*-shell rare-earth atoms are also of great interest.

2. Experimental

2.1. Synthesis

The starting materials for the synthesis of the $REPt_2Al_3$ and $RE_2Pt_3Al_4$ samples were pieces of the sublimed rare-earth elements (Y, Dy–Tm, and Lu from Smart Elements), platinum sheets (Agosi), and aluminum turnings (Koch Chemicals), all with stated purities better than 99.9%. For the $REPt_2Al_3$ compounds (*RE* = Y, Dy–Tm), the elements were weighed in the ideal 1:2:3 atomic ratio and arc-melted [19] in a water-cooled copper hearth under 800 mbar of argon pressure. The argon gas was purified with a titanium sponge (873 K), molecular sieves, and silica gel. Re-melting of the obtained buttons from each site several times enhanced the homogeneity. The as-cast buttons of the yttrium compound were crushed, and the fragments were sealed in quartz ampoules, placed in the water-cooled sample chamber of a high-frequency furnace (Typ TIG 5/300, Hüttinger Elektronik, Freiburg, Germany) [20], and heated until a softening of the piece was observed. The power was subsequently reduced by 10%, and the sample was kept at this temperature for 120 min before being cooled to room temperature. The other samples were annealed in muffle furnaces. They were heated to 1223 K and then kept at this temperature for 14 days, followed by slow cooling until they reached 573 K. Afterwards, the furnace was switched off. These different annealing procedures led to X-ray pure samples suitable for physical properties measurements. For the $RE_2Pt_3Al_4$ compounds (*RE* = Tm, Lu), the elements were weighed in the ideal 2:3:4 atomic ratio and arc-melted as described above. Again, an annealing step in a high-frequency furnace was subsequently conducted. The specimens are stable in air over weeks and show metallic luster; the ground samples are grey.

2.2. X-ray Image Plate Data and Data Collections

The polycrystalline samples were characterized at room temperature by powder X-ray diffraction on a Guinier camera (equipped with an image plate system, Fujifilm, Nakanuma, Japan, BAS-1800,) using Cu Kα_1 radiation and α-quartz (*a* = 491.30, *c* = 540.46 pm, Riedel-de-Haën, Seelze, Germany) as an internal standard. The lattice parameters (Table 1) were obtained from a least-squares fit. Proper indexing of the diffraction lines was ensured by an intensity calculation [21].

Irregularly shaped crystal fragments of the YPt_2Al_3 and $Lu_2Pt_3Al_4$ compounds were obtained from the annealed crushed buttons. The crystals were glued to quartz fibers using beeswax, and their quality was checked by Laue photographs on a Buerger camera (white molybdenum radiation, image plate technique, Fujifilm, Nakanuma, Japan, BAS-1800) for intensity data collection. The datasets were collected on a Stoe StadiVari four-circle diffractometer (Mo-Kα radiation (λ = 71.073 pm); μ-source; oscillation mode; hybrid-pixel-sensor, *Dectris Pilatus* 100 K [22]) with an open *Eulerian* cradle setup. Numerical absorption correction along with scaling was applied to the datasets. All relevant crystallographic data, deposition, and details of the data collection and evaluation are listed in Tables 2–8. Further details of the crystal structure investigation may be obtained from the Fachinformationszentrum Karlsruhe, 76344 Eggenstein-Leopoldshafen, Germany (Fax: +49-7247-808-666; E-Mail: crysdata@fiz-karlsruhe.de, http://www.fiz-karlsruhe.de/request_ for_deposited_data.html) by quoting the depository numbers CSD-434174 (YPt_2Al_3) and CSD-434175 ($Lu_2Pt_3Al_4$).

2.3. Energy Dispersive X-ray Spectroscopy (EDX) Data

The crystals measured on the diffractometer were analyzed semi-quantitatively using a Zeiss EVO MA10 scanning electron microscope with YF_3, TmF_3, LuF_3, Pt, and Al_2O_3 as standards. No impurity elements heavier than sodium (the detection limit of the instrument) were observed. The experimentally determined element ratios (YPt_2Al_3: 18 ± 2 at.% Y: 29 ± 2 at.% Pt: 53 ± 2 at.% Al; and $Lu_2Pt_3Al_4$: 20 ± 2 at.% Y: 36 ± 2 at.% Pt: 44 ± 2 at.% Al) were in close agreement with the ideal compositions (16.7:33.3:50 and 22.2:33.3:44.5), respectively. The deviations resulted from the irregular shape of the crystal surfaces (conchoidal fracture). Additionally, polycrystalline pieces from the annealed arc-melted buttons were embedded in a methylmethacrylate matrix and polished with diamond and SiO_2 emulsions of different particle sizes. During the first attempts to synthesize $TmPt_2Al_3$ and $LuPt_2Al_3$, phase segregation was observed; the secondary phases had the compositions $Tm_2Pt_3Al_4$ and $Lu_2Pt_3Al_4$.

2.4. Magnetic Properties Measurements

Fragments of the annealed buttons of the X-ray pure $REPt_2Al_3$ phases were attached to the sample holder rod of a vibrating sample magnetometer (VSM) unit using Kapton foil for measuring the magnetization $M(T, H)$ in a Quantum Design physical property measurement system (PPMS). The samples were investigated in the temperature range of 2.5–300 K with external magnetic fields up to 80 kOe. The magnetic data are summarized in Table 9.

2.5. X-ray Photoelectron Spectroscopy (XPS)

XPS was performed using an ESCALAB 250 Xi instrument (Thermo Fisher, East Grinsted, UK) with mono-chromatized Al Kα ($h\nu = 1486.6$ eV) radiation. All samples were cleaned by Ar^+ sputtering (MAGCIS ion gun, 36 keV) for 60 s to remove adventitious carbon. High-resolution spectra were measured with pass energies of 10 eV (Pt 4f, Al 2s, Al 2p, and C 1s) and 20 eV (Y 3d and Pr 3d). Peak deconvolution was performed using a Gaussian-Lorentzian peak shape by the software Avantage (Thermo Fisher). All spectra were referenced to remaining adventitious carbon at 284.8 eV. Because of the overlap of the Pt 4f and Al 2p signals, Al 2s was used for Al quantification. The obtained data are summarized in Table 10.

3. Results and Discussion

During attempts to synthesize aluminum intermetallics with the composition $REPt_2Al_3$, well-resolved X-ray powder patterns for the small rare-earth elements RE = Y, Dy–Tm were observed. For the thulium compound, additional reflections showed up in the unannealed sample, which were initially interpreted as impurities. Subsequently, single crystals from the yttrium sample were isolated and structurally investigated (*vide infra*). The large and early rare-earth elements (RE = La–Nd, Sm, Gd, and Tb) do not form the same structure type. Investigations on the structures formed by these elements are still ongoing. Attempts to synthesize $LuPt_2Al_3$ also yielded a diffraction pattern different from the slightly larger rare-earth elements Dy–Tm. As cast specimen, $TmPt_2Al_3$ and $LuPt_2Al_3$ were subsequently investigated by scanning electron microscopy coupled with energy dispersive X-ray spectroscopy (SEM/EDX). The impurity phase in $TmPt_2Al_3$ and the main phase in nominal $LuPt_2Al_3$ were found to be $Tm_2Pt_3Al_4$ and $Lu_2Pt_3Al_4$, respectively. Finally, samples with these compositions were prepared, and single crystals from $Lu_2Pt_3Al_4$ were isolated and investigated.

3.1. Structure Refinements

A careful analysis of the obtained intensity dataset of YPt_2Al_3 revealed an orthorhombic C-centered lattice. The centrosymmetric group *Cmmm* was found to be correct during structure refinement. A systematic check of the Pearson database [3], using Pearson code *oC*48 and Wyckoff sequence q^2pji^2hedb, gave no matches; hence, YPt_2Al_3 must be considered a new structure type. The starting atomic parameters were obtained using SuperFlip [23], implemented in Jana2006 [24,25]. The structure was refined on F^2 with anisotropic displacement parameters for all atoms. As a check for the correct composition and site assignment, the occupancy parameters were refined in a separate series of least-squares cycles. All sites were fully occupied within three standard deviations. No significant residual peaks were evident in the final difference Fourier syntheses. At the end, the positional parameters were transformed to the setting required for the group-subgroup scheme discussed below. Figure 1 depicts the X-ray powder diffraction pattern of YPt_2Al_3 along with the calculated pattern obtained using the positional information from the single-crystal structure refinement.

$Lu_2Pt_3Al_4$ was also found to crystallize in the orthorhombic crystal system with space group *Pnma*. A comparison with the Pearson database [3], using Pearson code *oP*36 and Wyckoff sequence c^9, indicated isotypism with $Ce_2Ir_3Sb_4$ [26,27]. The structure was refined on F^2 with anisotropic displacement parameters for all atoms. As a check for the correct composition and site assignment, the occupancy parameters were refined in a separate series of least-squares cycles. All sites were fully occupied within three standard deviations. No significant residual peaks were evident in the final difference Fourier syntheses. In the powder X-ray diffraction experiments, trace amounts of TmPtAl or LuPtAl (TiNiSi-type) were evident. Thermal treatment was not able to remove these impurities. The details of the structure refinement, final positional parameters, and interatomic distances are listed in Tables 2–8.

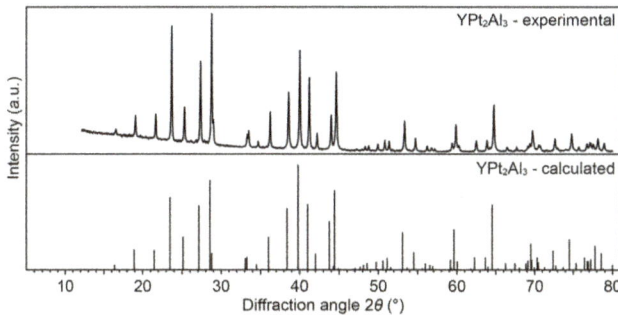

Figure 1. Experimental (**top**) and calculated (**bottom**) Guinier powder pattern (CuK$_{\alpha 1}$ radiation) of YPt_2Al_3.

Table 1. Lattice parameters of the orthorhombic $REPt_2Al_3$ series (YPt_2Al_3-type, rare-earth (*RE*) = Y, Dy–Tm), space group *Cmmm*, and $RE_2Pt_3Al_4$ series ($Ce_2Ir_3Sb_4$-type, *RE* = Y, Dy–Tm), space group *Pnma*.

Compound	*a* (pm)	*b* (pm)	*c* (pm)	*V* (nm³)
YPt_2Al_3	1080.73(6)	1871.96(9)	413.04(2)	0.8356
$DyPt_2Al_3$	1081.3(1)	1872.7(2)	413.93(5)	0.8382
$HoPt_2Al_3$	1079.26(4)	1869.46(6)	413.55(2)	0.8344
$ErPt_2Al_3$	1077.31(6)	1866.0(1)	413.14(4)	0.8305
$TmPt_2Al_3$	1075.38(9)	1862.6(1)	412.87(4)	0.8270
$Tm_2Pt_3Al_4$	1349.9(3)	418.22(8)	1143.7(2)	0.6429
$Lu_2Pt_3Al_4$	1343.4(2)	416.41(8)	1141.1(2)	0.6383

Table 2. Crystallographic data and structure refinement for YPt_2Al_3, space group $Cmmm$, $Z = 8$, own type and $Lu_2Pt_3Al_4$, space group $Pnma$, $Z = 4$, $Ce_2Ir_3Sb_4$-type.

Compound	YPt_2Al_3	$Lu_2Pt_3Al_4$
Molar mass, g mol^{-1}	560.0	1043.1
Density calc., g cm^{-3}	8.93	10.91
Crystal size, µm	25 × 40 × 55	30 × 30 × 40
Detector distance, mm	40	40
Exposure time, s	25	50
Integr. param. A, B, EMS	6.2; −5.2; 0.017	5.0; −4.1; 0.012
Range in hkl	±16; ±28, ±6	±21; ±6, ±18
θ_{min}, θ_{max}, deg	2.2–32.9	2.3–35.5
Linear absorption coeff., mm^{-1}	81.2	97.0
No. of reflections	11,714	21,601
R_{int}/R_σ	0.1124/0.0178	0.1411/0.1152
No. of independent reflections	942	1605
Reflections used [$I \geq 3\sigma(I)$]	795	679
$F(000)$, e	1872	1712
$R1/wR2$ for $I \geq 3\sigma(I)$	0.0341/0.0770	0.0415/0.0798
$R1/wR2$ for all data	0.0422/0.0780	0.1095/0.0940
Data/parameters	942/46	1605/56
Goodness-of-fit on F^2	2.22	1.23
Extinction coefficient	161(17)	73(6)
Diff. Fourier residues/e$^-$ Å^{-3}	−4.15/3.97	−4.98/4.51

Table 3. Atom positions and equivalent isotropic displacement parameters (pm^2) for YPt_2Al_3. U_{eq} is defined as one-third of the trace of the orthogonalized U_{ij} tensor.

Atom	Wyckoff Position	x	y	z	U_{eq}
Y1	2d	0	0	1/2	151(7)
Y2	4e	1/4	1/4	0	137(4)
Y3	2b	1/2	0	0	137(6)
Pt1	4h	0.27855(6)	0	1/2	120(2)
Pt2	8q	0.13928(4)	0.13927(3)	1/2	120(1)
Pt3	4i	0	0.33333(4)	0	136(2)
Al1	4j	0	0.2483(3)	1/2	128(14)
Al2	8q	0.3729(4)	0.1244(2)	1/2	138(10)
Al3	8p	0.2244(4)	0.0748(2)	0	160(11)
Al4	4i	0	0.1494(4)	0	157(16)

Table 4. Atom positions and equivalent isotropic displacement parameters (pm^2) for $Lu_2Pt_3Al_4$. U_{eq} is defined as one-third of the trace of the orthogonalized U_{ij} tensor. $y = 1/4$ all 4c.

Atom	x	z	U_{eq}
Lu1	0.01840(10)	0.71349(12)	199(3)
Lu2	0.29143(10)	0.57858(14)	218(3)
Pt1	0.13365(9)	0.24522(11)	196(3)
Pt2	0.38024(9)	0.06876(11)	201(3)
Pt3	0.62220(9)	0.58482(11)	189(3)
Al1	0.0017(7)	0.0827(9)	210(2)
Al2	0.0714(8)	0.4553(8)	180(20)
Al3	0.3017(7)	0.8651(9)	190(30)
Al4	0.3174(7)	0.2828(9)	170(20)

Table 5. Anisotropic displacement parameters (pm^2) for YPt$_2$Al$_3$. Coefficients U_{ij} of the anisotropic displacement factor tensor of the atoms are defined by: $-2\pi^2[(ha^*)^2U_{11}+ \ldots + 2hka^*b^*U_{12}]$. $U_{13} = U_{23} = 0$.

Atom	U_{11}	U_{22}	U_{33}	U_{12}
Y1	144(10)	139(11)	169(14)	0
Y2	137(7)	137(7)	136(9)	−1(6)
Y3	135(10)	140(10)	136(12)	0
Pt1	126(3)	122(3)	112(3)	0
Pt2	117(2)	131(2)	112(3)	7(1)
Pt3	147(3)	153(3)	107(4)	0
Al1	160(2)	120(20)	110(30)	0
Al2	138(16)	145(17)	130(20)	−18(14)
Al3	210(20)	156(18)	110(20)	40(15)
Al4	110(20)	220(30)	140(30)	0

Table 6. Anisotropic displacement parameters (pm^2) for Lu$_2$Pt$_3$Al$_4$. Coefficients U_{ij} of the anisotropic displacement factor tensor of the atoms are defined by: $-2\pi^2[(ha^*)^2U_{11} + \ldots + 2hka^*b^*U_{12}]$. $U_{13} = U_{23} = 0$.

Atom	U_{11}	U_{22}	U_{33}	U_{12}
Lu1	194(5)	209(6)	194(5)	−4(4)
Lu2	234(6)	228(6)	191(5)	−5(5)
Pt1	192(5)	206(6)	190(5)	−14(4)
Pt2	215(5)	203(5)	183(5)	−15(4)
Pt3	179(5)	204(5)	184(4)	−9(4)
Al1	200(40)	220(40)	210(40)	10(4)
Al2	280(50)	130(40)	120(30)	30(3)
Al3	120(40)	210(50)	240(50)	0
Al4	200(40)	130(40)	170(40)	20(3)

Table 7. Interatomic distances (pm) for YPt$_2$Al$_3$. All distances of the first coordination spheres are listed. All standard uncertainties were less than 0.2 pm.

Y1:	2	Pt1	300.7	Pt2:	1	Al1	253.3	Al2:	1	Pt2	253.7
	4	Pt2	300.7		1	Al2	253.7		1	Pt1	253.7
	4	Al4	347.3		2	Al4	256.0		2	Pt3	260.1
	8	Al3	347.6		2	Al3	256.0		1	Al2	274.4
					1	Pt1	300.7		1	Al1	274.8
Y2:	2	Pt3	311.7		1	Y1	300.7		2	Al3	277.3
	4	Pt2	315.8		1	Pt2	300.7		2	Y2	339.6
	2	Al3	328.8		1	Y2	315.8		2	Y3	339.9
	2	Al4	329.0								
	4	Al2	339.6	Pt3:	4	Al2	260.1	Al3:	2	Pt2	256.0
	4	Al1	339.7		2	Al1	260.5		2	Pt1	256.0
					2	Y2	311.7		2	Al2	277.3
Y3:	2	Pt3	311.7		1	Y3	311.7		1	Al4	279.6
	4	Pt2	315.8		2	Al3	343.6		1	Al3	279.7
	2	Al3	328.8		1	Al4	343.9		1	Y3	328.8
	2	Al4	329.0						1	Y2	328.8
	4	Al2	339.6	Al1:	2	Pt2	253.3		1	Pt3	343.6
	4	Al1	339.7		2	Pt3	260.5		2	Y1	347.6
					2	Al2	274.8				
Pt1:	2	Al2	253.9		2	Al4	277.1	Al4:	4	Pt2	256.0
	4	Al3	256.0		4	Y2	339.7		2	Al1	277.1
	2	Pt2	300.7						2	Al3	279.6
	1	Y1	300.7						2	Y2	329.0
	2	Y3	315.8						1	Pt3	343.9
									2	Y1	347.6

Table 8. Interatomic distances (pm) for $Lu_2Pt_3Al_4$. All distances of the first coordination spheres are listed. All standard uncertainties were less than 0.2 pm.

Lu1:	2	Pt3	298.8	Pt2:	2	Al2	253.1	Al2:	2	Pt2	253.1
	2	Pt1	302.0		1	Al4	254.9		1	Pt3	253.6
	2	Pt2	310.4		1	Al2	258.1		1	Pt2	258.1
	1	Al4	326.8		1	Al1	258.1		2	Al4	287.6
	1	Al2	327.1		2	Lu2	298.0		1	Lu2	302.7
	1	Al3	336.8		2	Lu1	310.4		2	Lu2	307.6
	1	Al1	338.9						1	Lu1	327.1
	2	Al4	343.4	Pt3:	1	Al1	250.3				
	2	Al1	344.4		1	Al2	253.5	Al3:	1	Pt1	250.2
	2	Al3	346.8		1	Al3	256.3		1	Pt3	256.3
					2	Al4	266.3		2	Pt1	266.0
Lu2:	1	Pt1	268.8		2	Lu2	294.9		2	Al3	280.5
	2	Pt3	294.9		2	Lu1	298.8		1	Al1	291.0
	2	Pt2	298.0						2	Lu2	312.7
	1	Al2	302.7	Al1:	1	Pt3	250.3		1	Lu1	336.8
	1	Al4	304.4		1	Pt2	258.1		2	Lu1	346.8
	1	Al2	307.6		2	Pt1	269.1				
	2	Al3	312.7		1	Al4	278.3	Al4:	1	Pt1	247.6
	2	Al1	312.9		1	Al3	291.0		1	Pt2	254.9
					2	Lu2	312.9		2	Pt3	263.3
Pt1:	1	Al4	247.6		1	Lu1	338.9		2	Al1	278.3
	1	Al3	250.2		2	Lu1	344.4		2	Al2	287.6
	2	Al3	266.0						1	Lu2	304.4
	1	Lu2	268.8						2	Lu1	326.8
	2	Al1	269.1						1	Lu1	343.4
	2	Lu1	302.0								

3.2. The YPt_2Al_3-Type Structure: Crystal Chemistry and Group-Subgroup Relations

The isostructural aluminum compounds of the $REPt_2Al_3$ series (RE = Y, Dy–Tm) crystallize in the orthorhombic crystal system, space group $Cmmm$, Pearson code $oC48$ and Wyckoff sequence q^2pji^2hedb. The lattice parameters (Figure 2) and unit cell volumes (Table 1) decrease from the dysprosium to the thulium compound, as expected, from the lanthanide contraction. The lattice parameters of the yttrium compound are in the same range, explainable by the similar ionic radii (Y^{3+}: 90 pm; Dy^{3+}: 91 pm; Ho^{3+}: 90 pm [28]).

As YPt_2Al_3 was investigated by single-crystal X-ray diffraction experiments, its crystal structure will be used for the structural discussion. A view of the crystal structure along the crystallographic c axis is depicted in Figure 3. The crystal structure features a polyanionic $[Pt_2Al_3]^{\delta-}$ network and shows full Pt/Al ordering. The heteroatomic Pt–Al distance range from 253 to 261 pm indicates substantial Pt–Al bonding, because these distances are in the range of the sum of the covalent radii for Pt+Al of 129 + 125 = 254 pm [29]. The polyanionic networks of YPtAl (TiNiSi-type) [30] and $Y_4Pt_9Al_{24}$ ($Y_4Pt_9Al_{24}$-type) [31] show similar distances of 257–269 and 246–274 pm, respectively. Additionally, homoatomic Al–Al distances ranging from 274 to 280 pm, and Pt–Pt distances of 301 pm can be found. The latter distances are slightly longer compared to what is found in elemental Pt (Cu-type, 284 pm) [32], while the aluminum distances are in line with elemental Al (Cu-type, 286 pm) [33]. Three crystallographically distinct Y^{3+} cations can be found in the cavities of the polyanion. They exhibit 18-fold coordination environments in the shape of six-fold-capped hexagonal prisms (Figure 4). The hexagonal prisms have slightly different compositions of $Y1@[Al_{12}+Pt_6]$, $Y2@[Al_8Pt_4+Al_4Pt_2]$, and $Y3@[Al_8Pt_4+Al_4Pt_2]$. The Y–Pt distances range from 301 to 316 pm; the Y–Al distances are 329 pm. While the former distances are in line with YPtAl, the latter distances are significantly longer (Y–Pt: 304–320 pm; Y–Al: 287–305 pm) [30].

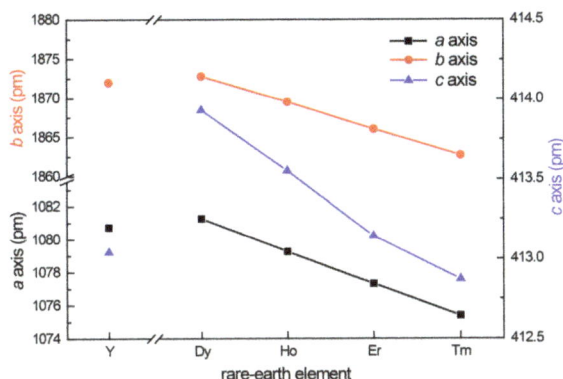

Figure 2. Plot of the unit cell parameters of the $REPt_2Al_3$ phases as a function of the rare-earth element.

A view of the unit cell along the c axis readily reminds us of the ternary $CaCu_5$-type derivatives $PrNi_2Al_3$ [15], YNi_2Al_3 [16], $DyNi_4Si$ [34], $CeCo_3B_2$ [14], and the recently found i7 superstructure of $PrNi_2Al_3$ [17]. Recoloring in intermetallics is found quite frequently, often accompanied by distortions and puckering within the respective structures [35]. These structural effects between different structure types can be investigated by so-called group-subgroup relations. The structures of $PrNi_2Al_3$ and YPt_2Al_3 are related by such a group-subgroup scheme, which is presented in the *Bärnighausen* formalism [36–39] in Figure 5. In the first step, an isomorphic symmetry reduction of index 4 takes place, which causes a doubling of the a and b axis, along with a splitting of the Pr ($1a$ to $1a$ and $3f$), Ni ($2c$ to $2c$ and $6l$), and Al ($3g$ to $6k$ and $6m$) sites. In the second step a *translationengleiche* transition of index 3 takes place, reducing the hexagonal symmetry from space group $P6/mmm$ to orthorhombic $Cmmm$. Again, a splitting of the crystallographic position occurs along with the introduction of additional degrees of freedom regarding the crystallographic positions. This enables a distortion of the polyanion and a recoloring of the crystallographic sites. The Y1 atoms finally occupy the $2d$ rather than the $2a$ site as suggested by the group-subgroup scheme. Hence, they are shifted by $1/2 z$ compared to the original position. The same shift is also observed in YNi_2Al_3 [16,35] and i7-$PrNi_2Al_3$ [17]. Refinement as orthorhombic trilling, as suggested by the *translationengleiche* symmetry reduction of index 3, is not necessary because the orthorhombic crystal system was found directly by the indexing routine.

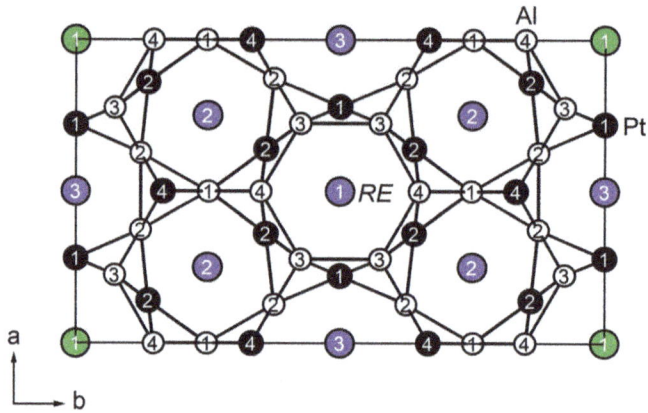

Figure 3. The crystal structure of YPt$_2$Al$_3$. Yttrium, platinum, and aluminum atoms are drawn as green/blue, black-filled, and open circles, respectively. The polyanionic [Pt$_2$Al$_3$]$^{\delta-}$ network is highlighted.

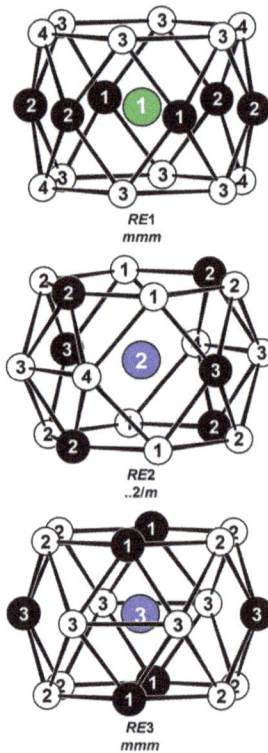

Figure 4. Coordination polyhedra surrounding the three crystallographically independent yttrium sites in YPt$_2$Al$_3$. Yttrium, platinum, and aluminum atoms are drawn as green/blue, black-filled, and open circles, respectively. The local site symmetries are given.

P6/mmm
PrNi₂Al₅ / CaCu₅

i4 2a,2b,c	void:1b 6/mmm	Pr:1a 6/mmm	Al:3g mmm	Ni:2c 6̄m2
	0	0	1/2	1/3
	0	0	0	2/3
	1/2	0	1/2	0

P6/mmm
t3, a, a+2b, c

RE1:1a 6/mmm	RE2:3f mmm	T1:6k m2m	X1:6m mm2	X2:6l mm2	T2:2c 6̄m2
0	1/2	~0.25	~0.25	~0.1667	1/3
0	1/2	0	2x	2x	2/3
0	0	1/2	1/2	0	0

Cmmm
YPt₂Al₅

calc

Y1:2d mmm	void:2a mmm	Y2:4e .2/m	Y3:2b mmm	Pt1:4h 2mm	Pt2:8q .m	Al1:4j m2m	Al2:8q .m	Al4:4i m2m	Al3:8p .m	Pt3:4i m2m
0	0	1/4	1/2	~0.25	~0.125	0	~0.375	0	~0.25	0
0	0	1/4	0	0	~0.125	~1/4	~0.125	~0.1667	~0.083	~0.333
1/2	0	0	0	1/2	1/2	1/2	1/2	0	0	0

refined

Y1:2d mmm	void:2a mmm	Y2:4e .2/m	Y3:2b mmm	Pt1:4h 2mm	Pt2:8q .m	Al1:4j m2m	Al2:8q .m	Al4:4i m2m	Al3:8p .m	Pt3:4i m2m
0	0	1/4	1/2	0.27855	0.13928	0	0.3729	0	0.2244	0
0	0	1/4	0	0	0.13927	0.2486	0.1244	0.1494	0.0748	0.33333
1/2	0	0	0	1/2	1/2	1/2	1/2	0	0	0

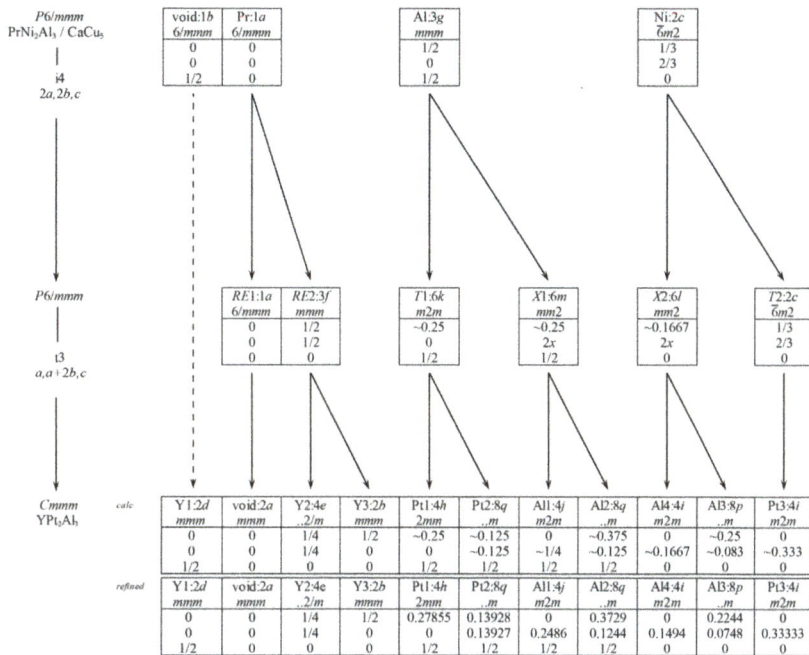

Figure 5. Group-subgroup scheme in the Bärnighausen formalism [36–39] for the structures of PrNi₂Al₃ and YPt₂Al₃. The index for the isomorphic (i) and *translationengleiche* (t) symmetry reduction, the unit cell transformation, and the evolution of the atomic parameters are given.

3.3. Crystal Chemistry of Tm₂Pt₃Al₄ and Lu₂Pt₃Al₄

Tm₂Pt₃Al₄ and Lu₂Pt₃Al₄ crystallize in the orthorhombic crystal system with space group *Pnma* (*oP*36, c^9) in the Ce₂Ir₃Sb₄-type structure [26,27]. In the following paragraph, Lu₂Pt₃Al₄ will be used for the structure description. As in the REPt₂Al₃ series, the platinum and aluminum atoms form a network. Figure 6 depicts the extended unit cell along [010], and the polyanionic [Pt₃Al₄]$^{6-}$ network and the two different lutetium sites are highlighted. The heteroatomic Pt–Al distances span a larger range (246–269 pm) compared to YPt₂Al₃; however, Pt–Al bonding is still present. In contrast to YPt₂Al₃, only additional Al–Al bonds can be found ranging from 278 to 300 pm. In the polyanion, no Pt–Pt bonds below 400 pm are found. The Al atoms form corrugated layers consisting of rectangles and hexagons in the boat conformation (Figure 7, top) that are capped by the Pt atoms (Figure 7, bottom).

The lutetium cations occupy two distinct crystallographic sites and are again found in the cavities of the polyanion. Lu1 is surrounded by 16 atoms in a four-fold-capped hexagonal prismatic environment (Lu1@[Al₆Pt₆+Al₄]; Figure 8, top), while Lu2 has a three-fold-capped pentagonal prismatic coordination sphere (Lu2@[Al₆Pt₄+Al₂Pt]; Figure 8, bottom). The Lu–Pt distances range from 299 to 310 pm, and the Lu–Al distances range from 327 to 347 pm. The Lu–Pt distances are in line with LuPtAl; the Lu–Al contacts are significantly longer (Lu–Pt: 302–327 pm; Lu–Al: 284–301 pm) [30].

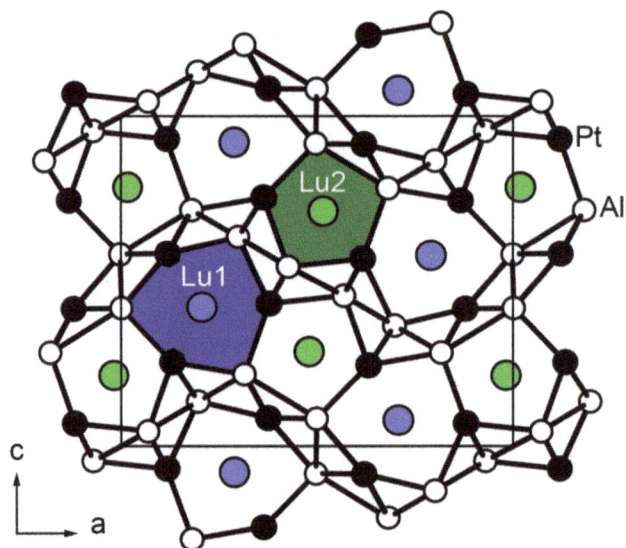

Figure 6. Extended crystal structure of $Lu_2Pt_3Al_4$ along [010]. Lutetium, platinum, and aluminum atoms are drawn as green/blue, black-filled, and open circles, respectively. The polyanionic $[Pt_3Al_4]^{\delta-}$ network and the two different coordination environments for the lutetium atoms are highlighted.

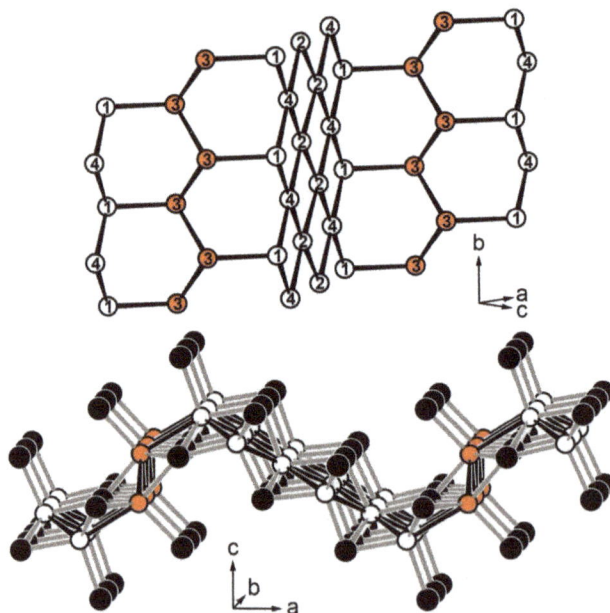

Figure 7. The Al arrangement in the crystal structure of $Lu_2Pt_3Al_4$ (top). The Pt atoms capping the layers are depicted in the bottom image. Platinum and aluminum atoms are drawn as black-filled and open circles, respectively. The Pt–Al bonds in the polyanionic $[Pt_3Al_4]^{\delta-}$ network are highlighted.

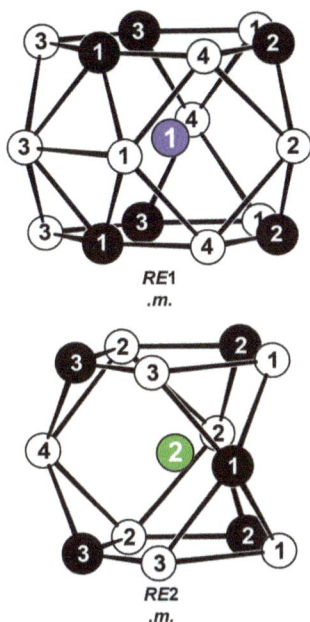

Figure 8. Coordination polyhedra surrounding the two crystallographically independent lutetium sites in $Lu_2Pt_3Al_4$. Lutetium, platinum, and aluminum atoms are drawn as green/blue, black-filled, and open circles, respectively. The local site symmetries are given.

3.4. Magnetic Properties

Magnetic susceptibility data has been obtained for the X-ray pure $REPt_2Al_3$ samples with $RE = Y$, Dy–Tm. The basic magnetic parameters that have been derived from these measurements are listed in Table 9. The temperature dependence of the magnetic susceptibility of the yttrium compound is depicted in Figure 9. YPt_2Al_3 is a *Pauli*-paramagnetic material with a room temperature susceptibility of $\chi = 1.85(1) \times 10^{-4}$ emu mol^{-1}. The weak upturn at lower temperature arises from small amounts of paramagnetic impurities. The present data clearly proves the absence of local moments on all constituent atoms. Thus, the magnetic properties of the remaining phases arise solely from the rare-earth elements.

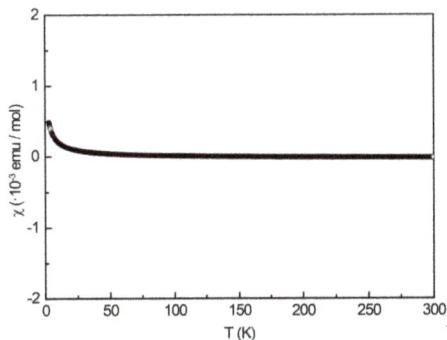

Figure 9. Temperature dependence of the magnetic susceptibility (data) of YPt_2Al_3 measured at 10 kOe.

The magnetic properties of $DyPt_2Al_3$, $HoPt_2Al_3$, $ErPt_2Al_3$, and $TmPt_2Al_3$ have been depicted in Figures 10–13. The top panels always depict the susceptibility and inverse susceptibility data (χ and χ^{-1}). The effective magnetic moments have been obtained from fitting the χ^{-1} data using the Curie–Weiss law between 50 and 300 K. They were calculated from the Curie constant according to $\mu_{eff} = \sqrt{\frac{3k_B C}{N_A}}$ [40,41]. All rare-earth atoms are in the trivalent oxidation state; the effective magnetic moments compare well within the calculated moments, as stated in Table 9. The calculated moments are tabulated [40,41] or can be calculated according to $\mu_{calc} = g\sqrt{J(J+1)}$ with $g = 1 + \frac{J(J+1)+S(S+1)-L(L-1)}{2J(J+1)}$ [40,41].

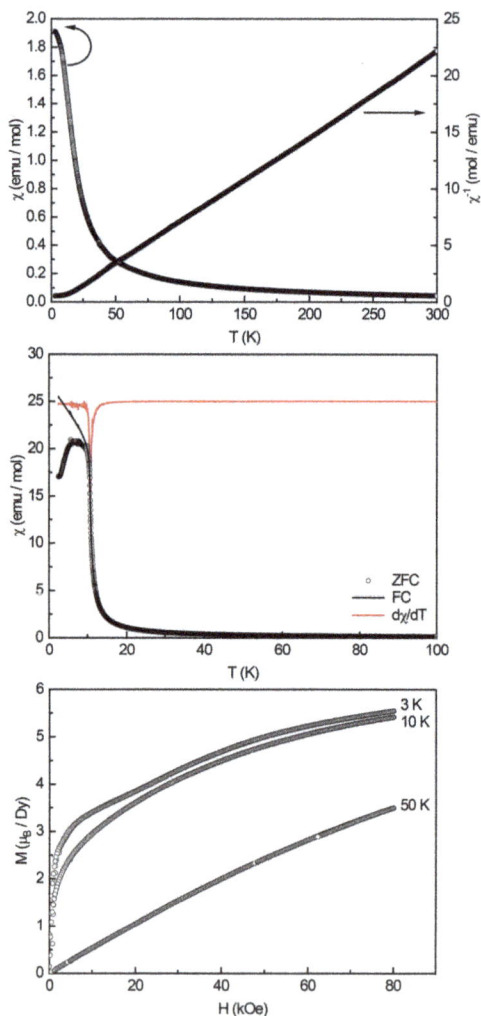

Figure 10. Magnetic properties of $DyPt_2Al_3$: (**top**) temperature dependence of the magnetic susceptibility χ and its inverse χ^{-1} measured at 10 kOe; (**middle**) zero-field-cooled/field-cooled (ZFC/FC) data (100 Oe) and the $d\chi/dT$ derivative (red curve) of the FC curve; and (**bottom**) magnetization isotherms recorded at 3, 10, and 50 K.

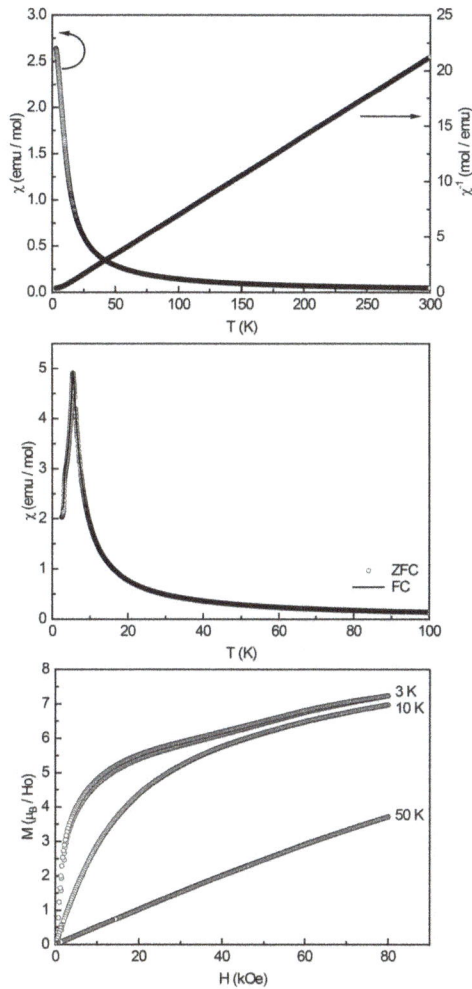

Figure 11. Magnetic properties of HoPt$_2$Al$_3$: (**top**) temperature dependence of the magnetic susceptibility χ and its inverse χ^{-1} measured at 10 kOe; (**middle**) zero-field-cooled/field-cooled (ZFC/FC) data (100 Oe); and (**bottom**) magnetization isotherms recorded at 3, 10, and 50 K.

Because a positive Weiss constant of θ_P is observed for the antiferromagnetically ordered compounds, the ordering phenomena could be a so-called Type-A antiferromagnetic ground state. In this ordered state, the intra-plane coupling is ferromagnetic while inter-plane coupling is antiferromagnetic [42]. From the zero-field-cooled/field-cooled (ZFC/FC) measurements depicted in the middle panels, it is evident that DyPt$_2$Al$_3$ and TmPt$_2$Al$_3$ exhibit ferromagnetic ordering at Curie temperatures of T_C = 10.8(1) and 4.7(1) K due to the plateau-like susceptibility at low temperatures. ErPt$_2$Al$_3$ exhibits no magnetic ordering down to 2.5 K, while HoPt$_2$Al$_3$ finally orders antiferromagnetically at T_N = 5.5(1) K, characterized by decreasing susceptibility below the Néel temperature. The Curie temperatures were obtained from the derivatives $d\chi/dT$ of the field-cooled curves (depicted in red) by determination of the temperature at the minimum in the derivative curve. The bottom panels finally display the magnetization isotherms measured at 3, 10, and 50 K. The 3 K isotherms of DyPt$_2$Al$_3$ and TmPt$_2$Al$_3$ show a fast increase at low fields, in line with the ferromagnetic ground state. The 3 K isotherm of HoPt$_2$Al$_3$ displays a slightly delayed increase, suggesting a spin-reorientation, in line with a weak antiferromagnetic ground state. The 3 K isotherm of DyPt$_2$Al$_3$ displays small 'wiggles', suggesting trace impurities, which are hardly noticeable in the ZFC/FC measurements. In the 3 K isotherm of HoPt$_2$Al$_3$, a small bifurcation is visible, also suggesting trace impurities, visible around 3 K in the ZFC/FC measurements. The isotherms at 50 K are all linear, in line with paramagnetic materials. The saturation magnetizations determined at 3 K and 80 kOe are all below the calculated values according to $g_J \times J$ (Table 9). The extracted values are, in all cases, lower than the expected moments, suggesting that the applied external field is not strong enough to achieve full parallel spin ordering.

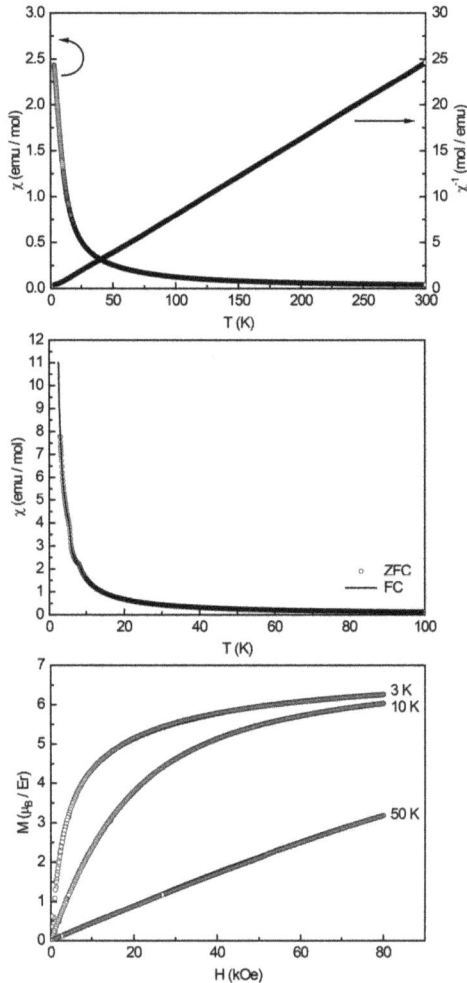

Figure 12. Magnetic properties of ErPt$_2$Al$_3$: (**top**) temperature dependence of the magnetic susceptibility χ and its inverse χ^{-1} measured at 10 kOe; (**middle**) zero-field-cooled/field-cooled (ZFC/FC) data (100 Oe); and (**bottom**) magnetization isotherms recorded at 3, 10, and 50 K.

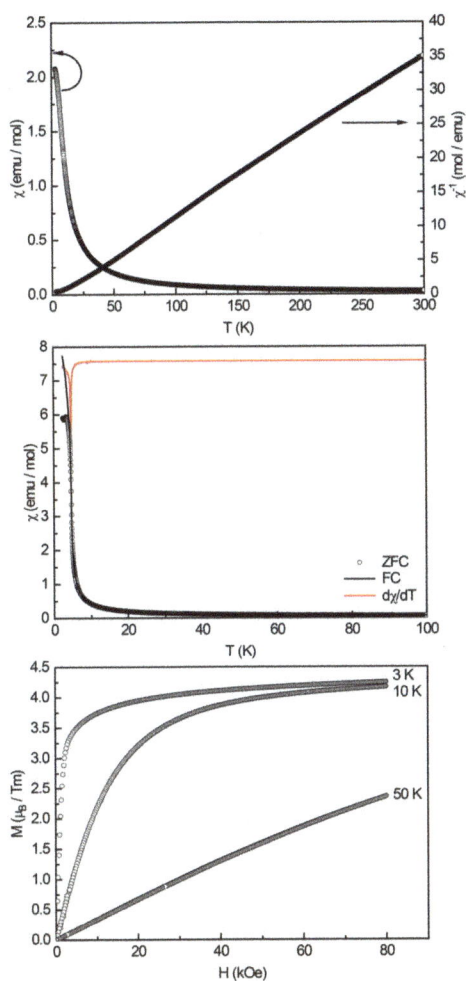

Figure 13. Magnetic properties of $TmPt_2Al_3$: (**top**) temperature dependence of the magnetic susceptibility χ and its inverse χ^{-1} measured at 10 kOe; (**middle**) zero-field-cooled/field-cooled (ZFC/FC) data (100 Oe) and the $d\chi/dT$ derivative (red curve) of the FC curve; and (**bottom**) magnetization isotherms recorded at 3, 10, and 50 K.

Table 9. Magnetic properties of the YPt_2Al_3-type compounds. T_N, Néel temperature; T_C, Curie temperature; μ_{eff}, effective magnetic moment; μ_{calc}, calculated magnetic moment; θ_p, paramagnetic Curie temperature; μ_{sat}, saturation moment; and saturation according to $g_J \times J$. The experimental saturation magnetizations were obtained at 3 K and 80 kOe.

	T_N (K)	T_C (K)	μ_{eff} (μ_B)	μ_{calc} (μ_B)	θ_p (K)	μ_{sat} (μ_B)	$g_J \times J$ (μ_B)
YPt_2Al_3		*Pauli*-paramagnetic χ(300 K) = 1.85(1) $\times$ 10^{-4} emu mol^{-1}					
$DyPt_2Al_3$	–	10.8(1)	10.67(1)	10.65	+1.0(1)	5.54(1)	10
$HoPt_2Al_3$	5.5(1)	–	10.59(1)	10.61	+2.0(1)	7.23(1)	10
$ErPt_2Al_3$	–	–	9.77(1)	9.58	+4.0(1)	6.26(1)	9
$TmPt_2Al_3$	–	4.7(1)	7.69(1)	7.56	+12.8(1)	4.25(1)	7

3.5. X-ray Photoelectron Spectroscopy

The reported compounds were described by rare-earth cations located in the cavities of a polyanion. Hence, the rare-earth atoms transfer electron density to the framework. This is in line with the effective magnetic moments of the rare-earth cations, proving them to be formally in a trivalent oxidation state. When looking at the electronegativities χ of the constituting elements of the $REPt_2Al_3$ series, it is evident that platinum is by far the most electronegative element. According to the *Pauling* scale, the values are as follows: $\chi(Y) = 1.22$, $\chi(Dy) = 1.22$, $\chi(Ho) = 1.23$, $\chi(Er) = 1.24$, $\chi(Tm) = 1.25$, $\chi(Pt) = 2.28$, and $\chi(Al) = 1.61$ [30]. Because all reported compounds are of a metallic nature, a distinct ionic platinide character as found in A_2Pt (A = K [43], Rb [43], Cs [43,44]) is highly unlikely, especially when considering the three-dimensional framework with strong covalent bonding character formed by Pt and Al. Therefore, XPS measurements were performed to investigate exemplarily the platinide character of YPt_2Al_3 along with the reference substances YPt_5Al_2 (*anti*-$ZrNi_2Al_5$-type [45]), YPtAl (TiNiSi-type [31]), and elemental Pt.

The obtained binding energies are listed in Table 10. Figure 14 depicts an exemplary fitted spectrum of YPt_2Al_3. As observed for $Ba_3Pt_4Al_4$ (E_b(Pt $4f_{7/2}$) = 70.9 eV) [46], the binding energies of YPt_2Al_3 (E_b(Pt $4f_{7/2}$) = 70.4 eV), YPt_5Al_2 (E_b(Pt $4f_{7/2}$) = 70.6 eV), and YPtAl (E_b(Pt $4f_{7/2}$) = 70.2 eV) are all shifted towards lower binding energies in comparison with elemental Pt (E_b(Pt $4f_{7/2}$) = 71.2 eV). This can be explained by a higher electron density at the Pt atoms, in line with an electron transfer from the less electronegative Y and Al atoms. The existing literature [46] shows shifts of the Pt $4f_{7/2}$ signal towards higher binding energies for the binary phases PtAl and $PtAl_2$ (PtAl: 71.6, $PtAl_2$: 72.1 eV), which can be explained by the bond formation between Pt and Al. In the ternary compounds, the additional electron transfer from the rare-earth atoms causes the lower binding energies and the 'platinide' character. While YPtAl and YPt_2Al_3 exhibit extensive Pt–Al bonding within the polyanion, only few heteroatomic Pt–Al bonds are observed in Pt-rich YPt_5Al_2. Consequently, the spectra of YPt_5Al_2 show the smallest shift in comparison with elemental Pt. In YPtAl, an equal ratio of Pt and Al can be found in contrast with YPt_2Al_3. In the latter compound, additional homoatomic bonding takes place; therefore, YPt_2Al_3 shows a smaller shift in the Pt $4f_{7/2}$ binding energies than YPtAl. As expected, Y is acting as electron donor, and therefore, the Y $3d_{5/2}$ signal is shifted by approximately 1 eV to higher binding energies (c.f. Table 10). However, all samples show a minor Y $3d_{5/2}$ component, that appears around 155.5 eV, in line with possible contaminations by traces of elemental yttrium.

Table 10. Fitted binding energies (in eV) determined by XPS of YPt_2Al_3, YPt_5Al_2, YPtAl, PrPtAl, and Pt and data from the literature. The determined uncertainty of binding energies in this work is ±0.1 eV.

Compound	Pt $4f_{7/2}$	Al $2s$	Y $3d_{5/2}$	Lit.
YPt_2Al_3	70.6	117.2	156.9	*
YPt_5Al_2	70.9	117.8	157.0	*
YPtAl	70.4	116.7	156.6	*
PrPtAl	70.7	**	–	*
Pt	71.4	–	–	*
Pt	71.2	–	–	[46]
$Ba_3Pt_4Al_4$	70.9	–	–	[46]
PtAl	71.6	–	–	[46]
$PtAl_2$	72.1	–	–	[46]

* This work. ** Signal invisible due to overlap with Pr $3d$.

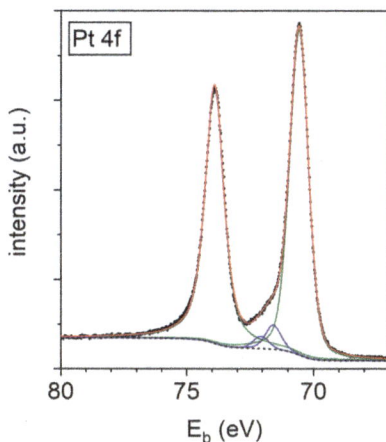

Figure 14. Fitted X-ray photoemission spectrum of Pt 4*f* in YPt$_2$Al$_3$. The experimental data is shown as black squares, the Pt 4*f* components are depicted in green, the Al 2*p* lines in blue, and the envelope function in red. The background is depicted as a dashed line.

4. Conclusions

Attempts to synthesize the CaCu$_5$-type related compounds *RE*Pt$_2$Al$_3$ with the late rare-earth elements Dy–Tm and Y led to the discovery of a new structure type, which was refined from single-crystal data obtained for YPt$_2$Al$_3$. The structure crystallizes in the orthorhombic space group *Cmmm* and can be derived from CaCu$_5$ by distortion and recoloring of the framework. Attempts to synthesize LuPt$_2$Al$_3$ led to the discovery of Lu$_2$Pt$_3$Al$_4$ (Ce$_2$Ir$_3$Sb$_4$-type), which was also refined from single-crystal data. The *RE*Pt$_2$Al$_3$ compounds could be obtained in phase pure form for property investigations. While YPt$_2$Al$_3$ is *Pauli*-paramagnetic, DyPt$_2$Al$_3$ to TmPt$_2$Al$_3$, in contrast, show paramagnetism in line with formal *RE*$^{3+}$ cations, along with magnetic ordering for *RE* = Dy, Ho, and Tm at low temperatures. Via XPS investigations, the binding energies of the constituent elements were investigated and compared with the electronegativities. In comparison with reference substances, the expected charge transfer onto the Pt atoms within the polyanionic [Pt$_2$Al$_3$]$^{\delta-}$ network could be proven.

Acknowledgments: We thank Dipl.-Ing. Ute Ch. Rodewald for its collection of the single-crystal diffractometer data. The XPS facility has been co-funded by the Deutsche Forschungsgemeinschaft (INST 184/144-1 FUGG).

Author Contributions: Fabian Eustermann and Simon Gausebeck performed the synthesis and the powder diffraction experiments; Fabian Eustermann, Simon Gausebeck and Oliver Janka solved and refined the single crystal structures; Carsten Dosche and Mareike Haensch measured and analyzed the XPS spectra; Oliver Janka measured and analyzed the magnetic data. Carsten Dosche, Gunther Wittstock and Oliver Janka wrote the paper.

Conflicts of Interest: The authors declare no conflict of interest.

References

1. Dshemuchadse, J.; Steurer, W. *Intermetallics: Structures, Properties, and Statistics*; Union of Crystallography, Oxford University Press: Oxford, UK, 2016.
2. Pöttgen, R.; Johrendt, D. *Intermetallics—Synthesis, Structure, Function*; De Gruyter: Berlin, Germany; Boston, MA, USA, 2014.
3. Villars, P.; Cenzual, K. *Pearson's Crystal Data: Crystal Structure Database for Inorganic Compounds (on DVD)*; Villars, P., Cenzual, K., Eds.; ASM International®: Materials Park, OH, USA, Release 2017/2018.
4. Friauf, J.B. The crystal structures of two intermetallic compounds. *J. Am. Chem. Soc.* **1927**, *49*, 3107–3114. [CrossRef]

5. Lieser, K.H.; Witte, H. Untersuchungen in den ternären Systemen Magnesium-Kupfer-Zink, Magnesium-Nickel-Zink und Magnesium-Kupfer-Nickel. *Z. Metallkd.* **1952**, *43*, 396–401.

6. Johanssonm, C.H.; Linde, J.O. Röntgenographische Bestimmung der Atomanordnung in den Mischkristallreihen Au-Cu und Pd-Cu. *Annal. Physik* **1925**, *78*, 439–460. [CrossRef]

7. Nowotny, H.N. Die Kristallstrukturen von Ni_5Ce, Ni_5La, Ni_5Ca, Cu_5La, Cu_5Ca, Zn_5La, Zn_5Ca, Ni_2Ce, MgCe, MgLa und MgSr. *Z. Metallkd.* **1942**, *34*, 247–253.

8. Ban, Z.; Sikirica, M. The crystal structure of ternary silicides ThM_2Si_2 (M = Cr, Mn, Fe, Co, Ni and Cu). *Acta Crystallogr.* **1965**, *18*, 594–599. [CrossRef]

9. Markiv, V.Y.; Matushevskaya, N.F.; Rozum, S.N.; Kuz'ma, Y.B. Investigation of aluminum-rich alloys of the system Zr-Ni-Al. *Inorg. Mater.* **1966**, *2*, 1356–1359.

10. Gupta, S.; Suresh, K.G. Review on magnetic and related properties of *RTX* compounds. *J. Alloy Compd.* **2015**, *618*, 562–606. [CrossRef]

11. Pöttgen, R.; Chevalier, B. Cerium intermetallics with ZrNiAl-type structure—A review. *Z. Naturforsch. B* **2015**, *70*, 289–304. [CrossRef]

12. Pöttgen, R.; Janka, O.; Chevalier, B. Cerium intermetallics CeTX—Review III. *Z. Naturforsch. B* **2016**, *71*, 165–191. [CrossRef]

13. Janka, O.; Niehaus, O.; Pöttgen, R.; Chevalier, B. Cerium intermetallics with TiNiSi-type structure. *Z. Naturforsch. B* **2016**, *71*, 737–764. [CrossRef]

14. Kuz'ma, Y.B.; Krypyakevych, P.I.; Bilonizhko, N.S. Crystal structure of $CeCo_3B_3$ and analogous compounds. *Dopov. Akad. Nauk Ukr. RSR* **1969**, *A31*, 939–941.

15. Rykhal, R.M.; Zarechnyuk, O.S.; Kuten, J.I. Isothermal section at 800 °C for the praseodymium-nickel-aluminium ternary system in the range of 0–33.3 at.% of praseodymium. *Dopov. Akad. Nauk Ukr. RSR* **1978**, *A10*, 1136–1138.

16. Zarechnyuk, O.S.; Rykhal, R.M. Crystal Structure of the Compound YNi_2Al_3 and Related Phases. *Visn. Lviv. Derzh. Univ., Ser. Khim.* **1981**, *23*, 45–47.

17. Eustermann, F.; Hoffmann, R.-D.; Janka, O. Superstructure formation in $PrNi_2Al_3$ and $ErPd_2Al_3$. *Z. Kristallogr.* **2017**, *232*, 573–581. [CrossRef]

18. Blazina, Z.; Westwood, S.M. On the structural and magnetic properties of the $CePt_{5-x}Al_x$ system. *J. Alloy Compd.* **1993**, *201*, 151–155. [CrossRef]

19. Pöttgen, R.; Gulden, T.; Simon, A. Miniaturisierte Lichtbogenapperatur für den Laborbedarf. *GIT Labor-Fachz.* **1999**, *43*, 133–136.

20. Pöttgen, R.; Lang, A.; Hoffmann, R.-D.; Künnen, B.; Kotzyba, G.; Müllmann, R.; Mosel, B.D.; Rosenhahn, C. The stannides YbPtSn and $Yb_2Pt_3Sn_5$. *Z. Kristallogr.* **1999**, *214*, 143–150. [CrossRef]

21. Yvon, K.; Jeitschko, W.; Parthé, E. LAZY PULVERIX, a computer program, for calculating X-ray and neutron diffraction powder patterns. *J. Appl. Crystallogr.* **1977**, *10*, 73–74. [CrossRef]

22. Dectris. *Technical Specification and Operating Procedure Pilatus 100K-S Detector System, Version 1.7*; Dectris: Baden-Daettwil, Switzerland, 2011.

23. Palatinus, L.; Chapuis, G. SUPERFLIP—A computer program for the solution of crystal structures by charge flipping in arbitrary dimensions. *J. Appl. Crystallogr.* **2007**, *40*, 786–790. [CrossRef]

24. Petříček, V.; Dušek, M.; Palatinus, L. *Jana2006. The Crystallographic Computing System*; Institute of Physics: Praha, Czech Republic, 2006.

25. Petříček, V.; Dušek, M.; Palatinus, L. Crystallographic Computing System JANA2006: General features. *Z. Kristallogr.* **2014**, *229*, 345–352. [CrossRef]

26. Schäfer, K.; Hermes, W.; Rodewald, U.C.; Hoffmann, R.-D.; Pöttgen, R. Ternary Antimonides $RE_2Ir_3Sb_4$ (*RE* = La, Ce, Pr, Nd). *Z. Naturforsch. B* **2011**, *66*, 777–783. [CrossRef]

27. Cardoso-Gil, R.; Caroca-Canales, N.; Budnyk, S.; Schnelle, W. Crystal structure, chemical bonding and magnetic properties of the new antimonides $Ce_2Ir_3Sb_4$, $La_2Ir_3Sb_4$, and $Ce_2Rh_3Sb_4$. *Z. Kristallogr.* **2011**, *226*, 657–666. [CrossRef]

28. Shannon, R.D. Revised effective ionic radii and systematic studies of interatomic distances in halides and chalcogenides. *Acta Crystallogr.* **1976**, *A32*, 751. [CrossRef]

29. Emsley, J. *The Elements*; Clarendon Press-Oxford University Press: Oxford, UK; New York, NY, USA, 1998.

30. Dwight, A.E. Crystal structure of equiatomic ternary compounds: Lanthanide-transition metal aluminides. *J. Less-Common Met.* **1984**, *102*, L9–L13. [CrossRef]

31. Thiede, V.M.T.; Fehrmann, B.; Jeitschko, W. Ternary Rare Earth Metal Palladium and Platinum Aluminides $R_4Pd_9Al_{24}$ and $R_4Pt_9Al_{24}$. *Z. Anorg. Allg. Chem.* **1999**, *625*, 1417–1425. [CrossRef]

32. Hull, A.W. The positions of atoms in metals. *Proc. Am. Inst. Electr. Eng.* **1919**, *38*, 1445–1466. [CrossRef]

33. Hull, A.W. A New Method of X-Ray Crystal Analysis. *Phys. Rev.* **1917**, *10*, 661–696. [CrossRef]

34. Morozkin, A.V.; Knotko, A.V.; Yapaskurt, V.O.; Yuan, F.; Mozharivskyj, Y.; Nirmala, R. New orthorhombic derivative of $CaCu_5$-type structure: RNi_4Si compounds (R = Y, La, Ce, Sm, Gd–Ho), crystal structure and some magnetic properties. *J. Solid State Chem.* **2013**, *208*, 9–13. [CrossRef]

35. Pöttgen, R. Coloring, Distortions, and Puckering in Selected Intermetallic Structures from the Perspective of Group-Subgroup Relations. *Z. Anorg. Allg. Chem.* **2014**, *640*, 869–891. [CrossRef]

36. Bärnighausen, H. Group-Subgroup Relations between Space Groups: A Useful Tool in Crystal Chemistry. *Commun. Math. Chem.* **1980**, *9*, 139–175.

37. Bärnighausen, H.; Müller, U. *Symmetriebeziehungen Zwischen den Raumgruppen als Hilfsmittel zur Straffen Darstellung von Strukturzusammenhängen in der Kristallchemie*; Universität Karlsruhe und Universität-Gh: Kassel, Germany, 1996.

38. Müller, U. Kristallographische Gruppe-Untergruppe-Beziehungen und ihre Anwendung in der Kristallchemie. *Z. Anorg. Allg. Chem.* **2004**, *630*, 1519–1537. [CrossRef]

39. Müller, U. *Symmetry Relationships between Crystal Structures*; Oxford University Press: New York, NY, USA, 2013.

40. Lueken, H. *Magnetochemie*; B.G. Teubner: Stuttgart/Leipzig, Germany, 1999.

41. Cheetham, A.K.; Day, P. *Solid State Chemistry Techniques*; Oxford University Press: New York, NY, USA, 1991.

42. Wollan, E.O.; Koehler, W.C. Neutron Diffraction Study of the Magnetic Properties of the Series of Perovskite-Type Compounds $[_{(1-x)}La_xCa]MnO_3$. *Phys. Rev.* **1955**, *100*, 545–563. [CrossRef]

43. Karpov, A.; Jansen, M. A New Family of Binary Layered Compounds of Platinum with Alkali Metals (A = K, Rb, Cs). *Z. Anorg. Allg. Chem.* **2006**, *632*, 84–90. [CrossRef]

44. Karpov, A.; Nuss, J.; Wedig, U.; Jansen, M. Cs_2Pt: A Platinide(-II) Exhibiting Complete Charge Separation. *Angew. Chem. Int. Ed.* **2003**, *42*, 4818–4821. [CrossRef] [PubMed]

45. Benndorf, C.; Stegemann, F.; Eckert, H.; Janka, O. New transition metal-rich rare-earth palladium/platinum aluminides with RET_5Al_2 composition: Structure, magnetism and ^{27}Al NMR spectroscopy. *Z. Naturforsch. B* **2015**, *70*, 101–110.

46. Stegemann, F.; Benndorf, C.; Bartsch, T.; Touzani, R.S.; Bartsch, M.; Zacharias, H.; Fokwa, B.P.T.; Eckert, H.; Janka, O. $Ba_3Pt_4Al_4$—Structure, Properties, and Theoretical and NMR Spectroscopic Investigations of a Complex Platinide Featuring Heterocubane $[Pt_4Al_4]$ Units. *Inorg. Chem.* **2015**, *54*, 10785–10793. [CrossRef] [PubMed]

crystals

MDPI

Review

The Crystal Orbital Hamilton Population (COHP) Method as a Tool to Visualize and Analyze Chemical Bonding in Intermetallic Compounds

Simon Steinberg [1] and Richard Dronskowski [1,2,*

[1] Institute of Inorganic Chemistry, RWTH Aachen University, D-52056 Aachen, Germany;
     simon.steinberg@ac.rwth-aachen.de
[2] Jülich-Aachen Research Alliance (JARA-FIT and -HPC), RWTH Aachen University,
     D-52056 Aachen, Germany
*   Correspondence: drons@HAL9000.ac.rwth-aachen.de; Tel.: +49-241-80-93642

Received: 25 April 2018; Accepted: 14 May 2018; Published: 18 May 2018

Abstract: Recognizing the bonding situations in chemical compounds is of fundamental interest for materials design because this very knowledge allows us to understand the sheer existence of a material and the structural arrangement of its constituting atoms. Since its definition 25 years ago, the Crystal Orbital Hamilton Population (COHP) method has been established as an efficient and reliable tool to extract the chemical-bonding information based on electronic-structure calculations of various quantum-chemical types. In this review, we present a brief introduction into the theoretical background of the COHP method and illustrate the latter by diverse applications, in particular by looking at representatives of the class of (polar) intermetallic compounds, usually considered as "black sheep" in the light of valence-electron counting schemes.

Keywords: COHP method; bonding analyses; intermetallic compounds

1. Introduction

The search for the inmost force which binds the world [1] and its constituents has generated enormous attention among scientists, and early theoretical research on the origin of this force identified a chemical power making atoms combine by so-called "valence bonds" [2]. The introduction of this very idea of valence bonds which arise from the atoms' valence electrons facilitated the developments of certain concepts, e.g., the octet rule [3], as fundamental relationships needed to rationalize the structural arrangements and the electronic structures of diverse molecules. For the case of solid-state materials, in particular ionic salts, first explorations to establish relationships between the structural arrangements and the electron counts of such solids employed empirical data and resulted in various solid-state rules which, for instance, were based on the ratios of ionic radii, the "strengths" of the electrostatic bonds, and the connectivities between diverse coordination polyhedra [4]. Further research on the distributions of the valence electrons in intermetallic compounds revealed additional notions, e.g., those first proposed by *Zintl* [5–10] and *Hume-Rothery* [11–13], respectively, to somehow correlate structural arrangements and the atoms' electronic nature. Even today, however, the existence of intermetallic compounds for which the electronic structures and, furthermore, the nature of bonding cannot be trivially categorized by applying one of the aforementioned concepts [14,15] underlines the need for different means in order to reveal the bonding nature in such materials.

To determine an unbiased picture of the nature of bonding in a given solid-state material, it seems necessary to extract the chemical bonding information from the electronic band structure that is computed for the respective material by means of a quantum-mechanical technique. For this purpose, the total energy of a many-particle system, i.e., the sum of the potential and kinetic energy, needs to be calculated solving the time-independent *Schrödinger* equation [16]

$$H\Psi = E\Psi \tag{1}$$

with H as the Hamiltonian operator, at least in principle. The wave function Ψ itself is then given by a properly symmetrized product of one-electron functions, so-called orbitals ψ. In the case of solid-state materials, the one-electron functions ψ are then constructed following *Bloch´s* theorem to incorporate the most important crystallographic symmetry: translation [17]. During the past nine decades, diverse techniques have been developed to approach the energies in order to construct the electronic band structure and densities-of-states of a given many-particle system [18]. The most successful trick to avoid calculating the many-electron wave function Ψ, however, is given by the framework of density-functional theory [19–21], the workhorse of numerous of today´s quantum-chemical computer programs. Here, the electron density, $\rho(r)$, plays a major role in generating an effective potential $V_{eff}(r)$ for non-existent pseudo-electrons plus some exchange-correlation correction term. In the end, one needs to solve the single-particle *Kohn–Sham* equations

$$\left[-\frac{1}{2}\nabla^2 + V_{eff}(r) \right] \psi_i(r) = E_i \psi_i(r) \tag{2}$$

and deals with one-electron ψ_i pseudo-functions to exactly cover the kinetic energy which is the largest contributor. One approach to eventually obtain the bonding information from the results of the electronic band structure calculations (regardless of wave-function or density-based) is the crystal orbital Hamilton population (COHP) method [22,23]. In the framework of the COHP technique, bonding, non-bonding, and antibonding interactions are identified for pairs of atoms (or orbitals) in a given solid-state material. In this review, we will present prototypical applications of the COHP procedure to solid-state materials. The outcome of this survey will demonstrate the efficiency of the method to reveal the actual nature of bonding and, furthermore, the causes for structural preferences in solids.

2. The COHP Method—An Introduction

Before demonstrating the applications of the COHP method for intermetallic compounds, we will provide a brief introduction into that technique. To obtain the energy of a many-particle system by solving the aforementioned *Schrödinger* or *Kohn–Sham* equations, one first needs to establish proper one-electron wave functions (orbitals). In the case of molecules consisting of a couple of atoms, these *molecular* orbitals are given by a linear combination of atomic orbitals (LCAO)

$$\psi_i(r) = \sum_A \sum_{\substack{\mu=1 \\ \mu \in A}}^{n} c_{\mu i} \phi_\mu(r), \tag{3}$$

composed of basis functions $\phi_\mu(r)$, i.e., the atomic orbitals of the atoms A, and the mixing coefficients $c_{\mu i}$. To obtain the energies of molecular systems, the Hamiltonian acts on the respective wave functions—but what is the best wave function among all possible functions? It is the one corresponding to the lowest total energy attainable by the variational principle [24]. Hence, the energy is differentiated with respect to the coefficients $c_{\mu i}$ ($\partial E / \partial c_{\mu i}$) leading to the secular determinant

$$\sum_A \sum_B \sum_\mu \sum_\nu c_{\mu i}^* c_{\nu i} H_{\mu\nu} - \sum_A \sum_B \sum_\mu \sum_\nu c_{\mu i}^* c_{\nu i} S_{\mu\nu} E_i = 0, \tag{4}$$

with $H_{\mu\nu} = \langle \phi_\mu | H | \phi_\nu \rangle$ and $S_{\mu\nu} = \langle \phi_\mu | \phi_\nu \rangle$ representing the on-site and off-site entries of the Hamiltonian and overlap matrices. In particular, $H_{\mu\mu}$ ($\mu = \nu$) contain the on-site or Coloumb integrals,

$H_{\mu v}$ ($\mu \neq v$) are the off-site, hopping, interaction or resonance integrals, and $S_{\mu v}$ represents the overlap integrals. Note that we use the *Dirac* bracket notation, in which ϕ_μ represent the conjugate complex wave functions, because ϕ can (and will be) a complex function including an imaginary part.

While the aforementioned procedure allows determining the energies of a given molecular system, at this point, one may wonder how the assignments of electrons to particular orbitals thereby providing information about the electron distribution can be accomplished. The numbers of electrons residing in the entire molecule and its molecular orbitals is given by

$$N = \sum_A \sum_{\substack{\mu \\ \mu \in A}} \sum_i^m f_i c_{\mu i}^2 + 2 \sum_A \sum_{B>A} \sum_{\substack{\mu \\ \mu \in A}} \sum_{\substack{v \\ v \in B}} \sum_i^m f_i c_{\mu i} c_{v i} S_{\mu v} \tag{5}$$

$$= \sum_A \sum_{\substack{\mu \\ \mu \in A}} \sum_i^m f_i \left(c_{\mu i}^2 + \sum_{B \neq A} \sum_{\substack{v \\ v \in B}} c_{\mu i} c_{v i} S_{\mu v} \right) \tag{6}$$

with f_i as the occupation numbers, which can be 0, 1, or 2. The first term of the Equation (5) represents the net populations, while the second one contains the overlap populations. Furthermore, the sum of both terms shown in the Equation (6) comprises the gross populations, following an early suggestion of *Mulliken*.

While the LCAO ansatz is used to solve *Schrödinger's* equation for molecules, the wave functions employed to solve *Schrödinger's* or the *Kohn–Sham* equations for solid-state materials are constructed based on *Bloch's* theorem [17]

$$\psi(\boldsymbol{k}, \boldsymbol{r} + \boldsymbol{T}) = e^{i\boldsymbol{k}\boldsymbol{T}}\psi(\boldsymbol{k}, \boldsymbol{r}), \tag{7}$$

with $\boldsymbol{T}$ as some lattice vector commutating with the Hamiltonian, $\psi(\boldsymbol{k}, \boldsymbol{r})$ as the crystal orbital at a specific site $\boldsymbol{r}$, and $\boldsymbol{k}$ as a new quantum number from reciprocal space. Without doubt, this is the most important theorem of theoretical solid-state science. The requirement of translational symmetry for constructing the wave functions fulfilling *Bloch's* theorem also demands that the crystal structure of a given material does not comprise any atomic sites showing positional of occupational disorders. Because the crystal structures of certain intermetallic compounds do comprise positionally or occupationally disordered sites (see Sections 3.2 and 3.3), it is necessary to examine the electronic structures of models *approximating* the actual crystal structures of such materials. To obtain numbers of electrons occupying the crystal orbitals in a given solid-state material, the $\boldsymbol{k}$-dependence of mixing coefficients $c_{\mu i}$ and $c_{v i}$ needs to be taken into consideration, leading to the $\boldsymbol{k}$-dependent density matrix

$$P_{\mu v}(\boldsymbol{k}) = \sum_i f_i c_{\mu i}^*(\boldsymbol{k}) c_{v i}(\boldsymbol{k}) \tag{8}$$

Inserting Equation (8) in the Equations (5) and (6), respectively, and taking $\int P_{\mu v}(\boldsymbol{k}) d\boldsymbol{k} = P_{\mu v}$ into account results in

$$N = \sum_A \sum_{\substack{\mu \\ \mu \in A}} P_{\mu \mu} + 2 \sum_A \sum_{B>A} \sum_{\substack{\mu \\ \mu \in A}} \sum_{\substack{v \\ v \in B}} Re\left[P_{\mu v} S_{\mu v}\right] \tag{9}$$

$$= \sum_{A} \sum_{\substack{\mu \\ \mu \in A}} \left(P_{\mu\mu} + \sum_{B \neq A} \sum_{\substack{\nu \\ \nu \in B}} Re[P_{\mu\nu}S_{\mu\nu}] \right), \tag{10}$$

with $Re[P_{\mu\nu}S_{\mu\nu}]$ as the real parts of the (possibly) complex off-diagonal entries. To get hold of the numbers of electrons N being dependent of the (band) energy, it is convenient to introduce the density-of-states matrix

$$P_{\mu\nu}(E) = \sum_{i} f_i c_{\mu i}^* c_{\nu i} \delta(\varepsilon - \varepsilon_i) \tag{11}$$

which is obtained from a differentiation of $P_{\mu\nu}$ with respect to E

$$P_{\mu\nu} = \int_{-\infty}^{\varepsilon_F} P_{\mu\nu}(E) dE \tag{12}$$

Combining the Equations (9) and (10) with Equation (11) yields

$$N = \int_{-\infty}^{\varepsilon_F} \sum_{A} \sum_{\substack{\mu \\ \mu \in A}} P_{\mu\mu}(E) dE + \int_{-\infty}^{\varepsilon_F} 2 \sum_{A} \sum_{B>A} \sum_{\substack{\mu \\ \mu \in A}} \sum_{\substack{\nu \\ \nu \in B}} Re[P_{\mu\nu}(E)S_{\mu\nu}] dE \tag{13}$$

$$= \int_{-\infty}^{\varepsilon_F} \sum_{A} \sum_{\substack{\mu \\ \mu \in A}} \left(P_{\mu\mu}(E) dE + \sum_{B \neq A} \sum_{\substack{\nu \\ \nu \in B}} Re[P_{\mu\nu}(E)S_{\mu\nu}] \right) dE \tag{14}$$

The (second) off-diagonal contribution of the gross population in Equation (14), i.e., $Re[P_{\mu\nu}(E)S_{\mu\nu}]$, comprises the overlap-population-weighted densities-of-states and, hence, provides the essential information regarding the distributions of the electrons *between* the atoms in any given solid-state material. This very technique to extract the bonding information based on the overlap-population-weighted densities-of-states has been dubbed as the crystal orbital overlap population (COOP) method [25]. In the framework of the COOP method, bonding and antibonding interactions are represented by positive and negative COOP values, respectively, while nonbonding interactions are indicated by zero COOPs. The COOP method has been largely employed for electronic-structure computations based on the semiempirical extended Hückel theory (EHT), which is principally equivalent to the empirical tight-binding approach including overlap [26]. In the EHT case, only the valence orbital are taken into account. In addition, certain entries of the secular determinant Equation (4), in particular, those of the interaction ($H_{\mu\nu}$) integrals, are parameterized; to do so, the interaction matrix elements are determined solving the *Wolfsberg–Helmholz* formula [27] that employs Coulomb matrix elements ($H_{\mu\mu}$) evolved from experimentally determined ionization potentials, while the overlap matrix elements are calculated utilizing *Slater*-type orbitals.

As denoted in the previous paragraph, theoretically determining the ground state of a given solid-state system can be carried out by different approaches, and the computational challenges arising from the notorious electron–electron interactions can be so tremendous that they easily overcome today´s (and also tomorrow's) available computational resources [18]. To nonetheless come to numerically powerful solutions of this problem, it has turned out extraordinarily convenient to concentrate on the electron density $\rho(r)$ which arises from effective one-electron wave functions (orbitals) $\psi_i(r)$ in the framework of density-functional-theory [19–21] (DFT), leading to

$$\rho(r) = \sum_{i=1}^{N} |\psi_i(r)|^2 \tag{15}$$

Using DFT jargon, the total energy E is a functional of the electron density ρ because the latter depends on the spatial coordinates r (i.e., a functional is a function which also depends on a function). To make DFT succeed, one sets up a pseudo-system of non-interacting electrons of the same density and adds a correction term for exchange and correlation. So, the effective potential $V_{eff}(r)$ employed in DFT computations (see Equation (2)) is the sum of the external potential $V_{ext}(r)$ (= the Coulomb potential from the nuclei), the so-called Hartree potential of the electrons $V_{Hartree}(r)$, and the exchange-correlation potential $V_{XC}(r)$, a correction term:

$$V_{eff}(r) = V_{ext}(r) + V_{Hartree}(r) + V_{XC}(r) \tag{16}$$

$$= V_{ext}(r) + \int \frac{\rho(r')}{|r - r'|} d^3 r' + \frac{\delta E_{XC}}{\delta\rho(r)} \tag{17}$$

Trivially, the accuracy of any DFT calculation mostly depends on the quality of $V_{XC}(r)$. As basis sets, solid-state people typically use plane waves and related functions fulfilling *Bloch's* theorem Equation (7); traditionally, methods based on cellular (augmentation) techniques to separate outer and inner parts of the individual atoms were used early on [18]. Particularly, these techniques are based on the concept to first solve *Schrödinger's* equation for a single atom within one cell and, subsequently, to glue the energy-dependent atomic functions (that is, the numerically derived atomic orbitals) together with those located on neighboring atoms. In this connection, it is convenient to define muffin-tin spheres being boundaries between the potentials in the regions close to the nuclei and the (zero) potentials in the interstitial regions. Accordingly, two-region potentials are obtained, and the wave functions which are constructed following the augmented-plane wave method [28] can be depicted as hybrid functions composed of atomic functions inside the muffin-tin spheres and single plane waves outside these spheres; however, the computations are slow because of the energy-dependence of the augmented plane waves, hence no simple diagonalization. To drastically enhance the speed of the computations, Taylor series of the energy for the radial parts of the functions within the spheres were developed and then truncated after the second term [29]. This approach (and its modifications) to compute the electronic structures in solid-state materials has been implemented in diverse quantum-mechanical techniques dubbed as *linear methods* including the (Full-potential) Linearized Augmented Plane-Wave, (F)LAPW, and Linearized Muffin-Tin Orbital, LMTO, methods. More recently, a related approach named Projector-Augmented Wave (PAW) method [30] has been introduced. In this (all-electron) technique, pseudopotentials (also dubbed effective core potentials) replace the ion-electron potentials and are optimized during the computations utilizing full wave functions. At this stage, one may wonder how the nature of covalent (hence, localized) bonding in solid-state materials may be interpreted based on the results of density-functional-theory-based computations, in particular for the reason that the use of essentially delocalized plane waves hinders the extractions of information regarding the bonding from the calculations [18]. We will come back to this point shortly.

Because extracting the bonding information from the electronic-structure computations strongly depends on the types of employed orbitals in the case of the aforementioned extended-Hückel based COOP method, applying this procedure to the results of various DFT computations with whatever kind of bases sets is rather problematic [22]. To nonetheless gain the bonding information from DFT, it is more appropriate to partition the band-structure energy leading to the Hamilton-matrix-weighted densities-of-states as shown in the following [22].

Under consideration of the secular determinant Equation (4), which can be transformed to

$$\sum_A \sum_B \sum_\mu \sum_\nu c_{\mu i}^* c_{\nu i} H_{\mu\nu} = \sum_A \sum_B \sum_\mu \sum_\nu c_{\mu i}^* c_{\nu i} S_{\mu\nu} E_i, \tag{18}$$

it is plausible that the overlap integrals used to define the overlap-population-weighted densities-of-states in the Equation (14) can be replaced by the Hamilton matrix elements; however, it should be noted that a straightforward transformation which might be implied by the relation in the Equation (18) cannot be made. Nonetheless, using an analogy to the aforementioned overlap-population-weighted densities-of-states, the densities-of-states matrix Equation (11) is weighted by the Hamilton matrix elements resulting in the Crystal Orbital Hamilton Populations (COHP), according to

$$\sum_A \sum_{\substack{\mu \\ \mu \in A}} \sum_B \sum_{\substack{\nu \\ \nu \in B}} H_{\mu\nu} \sum_i f_i c^*_{\mu i} c_{\nu i} \delta(\varepsilon - \varepsilon_i) = \sum_A \sum_{\substack{\mu \\ \mu \in A}} \sum_B \sum_{\substack{\nu \\ \nu \in B}} H_{\mu\nu} \, P_{\mu\nu}(E) = \sum_A \sum_{\substack{\mu \\ \mu \in A}} \sum_B \sum_{\substack{\nu \\ \nu \in B}} \text{COHP}_{\mu\nu}(E) \quad (19)$$

Because the off-site COHP terms (A ≠ B) originate from the interactions between the pairs of atoms in a given solid-state structure, these contributions are generally taken into consideration for the bonding analyses. More recently, a generalized variant of the COHP dubbed Density-of-Energy (DOE) function has been introduced [31]. In the DOE function, the (interatomic) off-site as well as the (atomic) on-site contributions are included such that the entire band energy can be obtained from the energy integral of the DOE function

$$E^{\text{band}} = \int_{-\infty}^{\varepsilon_F} \text{DOE}(E) dE. \quad (20)$$

In contrast to the COOP method, bonding and antibonding interactions are represented by negative and positive values of the COHP, respectively, while non-bonding interactions are indicated by zero values of the COHP. Somewhat simplified, a positive COOP denotes bonding, just like a negative COHP; in case one plots −COHP data, the shape of the two functions look similar, with bonding spikes to the right. Furthermore, a direct comparison between the COHP values of compounds with dissimilar compositions cannot be made, because the average electrostatic potential in each density-functional-theory-based computation is set to an arbitrary 'zero' energy, whose relative position can vary from system to system [18]. To identify the differences between the bonding situations in compounds with dissimilar compositions, the cumulative ICOHP/cell, i.e., the sum of the ICOHP/bond values of the nearest neighboring interactions weighted by the respective bond frequencies, are typically projected as the percentage contributions to the net bonding capacities (examples of this procedure have been provided in the Section 3).

The COHP technique has been first implemented in programs employing LMTO-based methods (see above) to compute the electronic structures of solid-state materials [22]. In a sense, LMTO theory introduced first-principles techniques to solid-state chemistry. This is because the COHP approach requires the use of crystal orbitals derived from *local* basis sets, the LMTO success ingredient. The more popular and essentially delocalized plane-wave-based computations of today, however, are blind for chemical-bonding analysis, so one first needs to reconstruct both Hamilton and overlap matrix elements using auxiliary atomic orbitals. In other words, such information may be *projected* from plane waves by means of the projected Crystal Orbital Hamilton Population (pCOHP) technique, a modern descendant of the COHP method [23]. The pCOHP technique has been implemented in the *Local Orbital Basis Suite Towards Electronic-Structure Reconstruction* (LOBSTER) program being available at www.cohp.de free of charge and compatible with a growing list of quantum-mechanical programs (VASP [32–36], ABINIT [37–40], Quantum ESPRESSO [41,42]) to extract the information regarding the nature of bonding from the electronic-structure computations [22,23,43,44].

After this brief introduction into the theoretical background of the COHP method, we will now present some applications for the examples of diverse intermetallic compounds to show the strengths of this technique in revealing the bonding situations in solid-state materials (Figure 1).

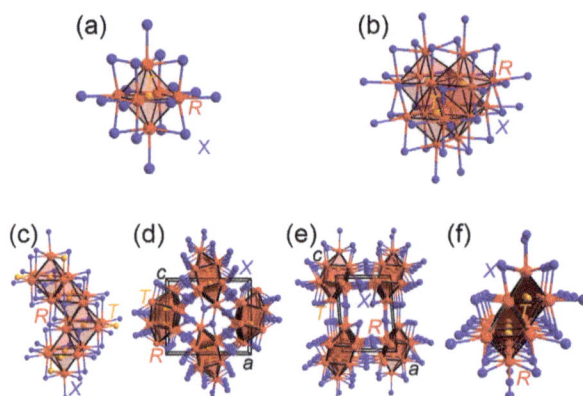

Figure 1. Representations of (**a**) isolated; (**b**) tetrameric; and (**c**−**f**) chains of transition-metal (T; yellow) centered rare-earth metal (R; red) cluster halides (X; blue). For instance, isolated clusters as shown in (**a**) have been identified for the crystal structure of the $[TR_6]X_{10}$ and $[TR_6]X_{12}R$ types of structure, whereas tetramers as presented in (**b**) have been observed for the rare-earth transition-metal halides crystallizing with the $[T_4R_{16}]X_{28}[R_4]$, $[T_4R_{16}]X_{20}$, $[T_4R_{16}]X_{24}[RX_3]_4$, and $[T_4R_{16}]X_{24}$ types of structure. To date, diverse types of structure which are composed of rare-earth cluster chains enclosing the endohedral transition-metal atoms have been determined to crystallize with the net formula $[TR_3]X_3$ shown in (**c**−**f**), i.e., (**c**) the cubic [PtPr$_3$]I$_3$-type, (**d**) the orthorhombic [RuPr$_3$]Cl$_3$-type, (**e**) the monoclinic [IrY$_3$]I$_3$-type, and (**f**) the monoclinic [RuPr$_3$]I$_3$-type. Adapted in part from reference [45].

3. Applications of the COHP Method to Intermetallic Compounds

3.1. Rare-Earth Transition-Metal Halides as Anti-Werner-Fashioned Complexes

Explorative research on the phase diagrams for the oxides and halides of the vanadium- to manganese-group elements revealed the existence of so-called "reduced" transition-metal halides and oxides, whose crystal structures comprise transition-metal building units of polyhedral forms dubbed as metal atom clusters [46–48]. In particular, the excess valence electrons of the transition metals, which do not participate in transition-metal−halide bonding, are distributed in transition-metal–transition-metal bonds formed by the d-orbitals of the metal atoms within the clusters [49]. Additional research on the phase diagrams for the halides of the scandium- to titanium-group elements being exceptionally poor in electron density identified halides composed of transition-metal clusters enclosing endohedral atoms, so-called 'interstitials' [48,50–55]. Furthermore, these rare-earth metal clusters encapsulating the interstitials have been observed to be surrounded solely by halide ligands (such types of clusters are typically classified as 'isolated'), or to share common vertices, edges and faces to assemble oligomers, chains, and sheets (Figure 1) [48,50–55]. As an outcome, these compounds comprise oxidized networks of metals surrounded by monoatomic ligands such that this group of materials has been assigned to the class of the polar intermetallic compounds [15].

To date, a broad variety of elements including transition-metal as well as main-group elements has been observed to be incorporated in the centers of the rare-earth clusters [54,55]. At that point, one may wonder how the electronic structure and the bonding situation typically expected for a transition-metal atom cluster is affected by the incorporation of an endohedral atom. In fact, the orbitals of the interstitials combine with the transition-metal cluster-based orbitals such that interstitial–transition-metal bonds evolve [49]. While the electronic structures and bonding motifs of isolated transition-metal clusters enclosing endohedral atoms can be described based on a molecular orbital ansatz [49], yet, the bonding situations of interstitially centered transition-metal cluster oligomers, chains, and sheets showing more metal–metal condensation and bonding need to be examined based on the electronic band structures of these materials. In the following, we will provide a survey of analyses employing the COHP method to determine the bonding motifs for the examples of the most prolific representatives of the transition-metal (T) centered rare-earth (R) cluster halides (X).

A large number of halides which are composed of isolated rare-earth clusters encapsulating transition-metal atoms has been identified to adopt the $[TR_6]X_{10}$ and $[TR_6]X_{12}R$ types of structure (note that the brackets define the components of the clusters) [56–58]. In particular, the endohedral transition-metal atoms which can be an element from the manganese- to zinc-groups for the $[TR_6]X_{12}R$-type and from the iron- to nickel-groups for the $[TR_6]X_{10}$-type are coordinated by octahedra constituted by the rare-earth atoms, while all edges of the octahedral $[TR_6]$ clusters are capped by halide ligands. The electronic structures and bonding situations in the isolated $[TR_6]$ clusters may be depicted by the aforementioned molecular orbital (MO) theory-based approach, for which the interstitial-based valence orbitals, i.e., the d-orbitals in the forms of t_{2g} and e_g sets, are combined with the fragment orbitals constituted by the rare-earth metal d-orbitals in the skeletons of the clusters (Figure 2) [56,57,59,60]. The results of the MO-theory-based computations on isolated empty and filled transition-metal clusters helped to substantiate electron-counting schemes which address the numbers of cluster-based electrons (CBEs) being available for the metal−metal bonds within the clusters. More specifically, the totals of the CBEs are usually obtained by subtracting the numbers of halide ligands from the sums of transition-metal valence electrons. While closed-shell configurations are achieved typically for counts of 16 CBEs in empty octahedral transition-metal clusters whose edges are capped by halide ligands [61], the outcome of the MO-theory-based examinations for the transition-metal-centered rare-earth clusters revealed that closed-shell configurations are accomplished for 18 CBEs (Figure 2) [56,57]. For instance, a closed-shell configuration is expected for the previously identified $[CoY_6]I_{12}Y$, for which an application of the electron-counting scheme yields a total of 18 CBEs ($=9 + 7·3 − 12$); however, the observed ranges of transition-metals being incorporated in the rare-earth clusters indicate certain electronic flexibilities. Chemical-bonding analysis based on the COHP curves of a number of $[TR_6]X_{10}$-type compounds demonstrated that the bond energy is optimized for a halide with a nickel-group interstitial in agreement with the outcome of the previous MO-theory-based calculations [58]. Furthermore, the bonding within the octahedral $[TR_6]$ clusters was shown to be dominated by the heteroatomic $T−R$ interactions besides much lesser $R−R$ bonding [58].

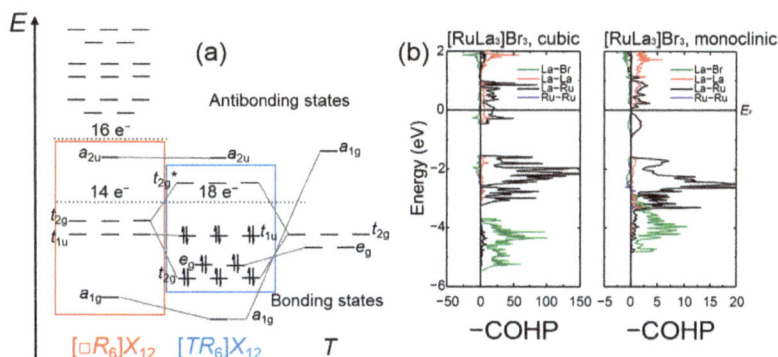

Figure 2. (**a**) Molecular-orbital scheme of an isolated transition-metal (T) centered rare-earth (R) cluster enclosed by halide ligands (X): the cluster-based orbitals (middle) are constructed by the valence orbitals of the endohedral transition-metal atom (right) and the fragment orbitals of the rare-earth skeleton (left). (**b**) COHP curves of the homoatomic and heteroatomic interactions for two structurally different representatives of the $[TR_3]X_3$-type, whose crystal structures comprises chains of transition-metal-centered rare-earth clusters (see main text): the Fermi levels, E_F, are represented by the black horizontal lines. Parts of that figure are adapted from reference [45].

Among the group of rare-earth transition-metal halides composed of cluster oligomers, the crystal structures of a large number of compounds have been identified to contain one particular type of tetramer, $[T_4R_{16}]$ (Figure 1). In particular, this sort of tetramer has been observed for the crystal structures of four different types of rare-earth transition-metal halides, that are, the $[T_4R_{16}]X_{28}[R_4]$ [62], $[T_4R_{16}]X_{20}$ [63], $[T_4R_{16}]X_{24}[RX_3]_4$ [64], and $[T_4R_{16}]X_{24}$ [64] types of structure. The tetramers consist of pairs of dimers, i.e., two transition-metal centered octahedral rare-earth clusters sharing one common edge, which are positioned perpendicular to each other and condensed via four joint edges. The $[T_4R_{16}]$ units are enclosed by 36 halide ligands, and their R_{16} fragments can also be depicted as all-vertices-truncated supertetrahedra or *Friauf*-polyhedra. An examination [65] of the hitherto identified rare-earth transition-metal halides whose crystal structures comprise such tetramers suggested that the maximum numbers of CBEs for intracluster bonding are attained for totals of 15 electrons per transition-metal centered rare-earth octahedron. Examinations [66] of the electronic structures for diverse halides composed of these tetramers bared that the Fermi levels fall into gaps corresponding to closed-shell configurations in halides for which CBE counts of 15 electrons per $[TR_6]$ cluster are achieved. An additional bonding analysis [66] based on the COHP curves and their integrated values (Table 1) indicates that the majority of the bonding interactions reside between the heteroatomic $R-T$ as well as $R-X$ separations, whereas the homoatomic $R-R$ and $T-T$ interactions play minor, but evident roles.

The group of rare-earth transition-metal halides with the net formula $[TR_3]X_3$ has emerged as a prolific class of compounds comprising cluster chains. To date, five different structure types have been identified for this group: the cubic [PtPr$_3$]I$_3$-type [67], the tetragonal [NiLa$_3$]Br$_3$-type [68], the orthorhombic [RuPr$_3$]Cl$_3$-type [69], and two independent monoclinic types of structure, i.e., the [RuPr$_3$]I$_3$ and [IrY$_3$]I$_3$ types of structure [70] (Figure 1). In particular, the monoclinic [RuPr$_3$]I$_3$-type has been observed solely for bromides and iodides which comprise early lanthanide clusters ($R = $ La$-$Pr) containing iron-group elements, while the cubic [PtPr$_3$]I$_3$-type has been identified for early lanthanide ($R = $ La$-$Pr) cluster bromides and iodides containing elements from the iron- to copper-groups as endohedral atoms [71]. In the crystal structure of the monoclinic [RuPr$_3$]I$_3$-type, the octahedral $[TR_6]$ clusters share four common edges with neighboring clusters forming bioctahedral chains, whereas the $[TR_6]$ octahedra in the cubic [PtPr$_3$]I$_3$-type structure condense via three common edges with other $[TR_6]$ clusters to helical chains. The existence of the cubic [PtPr$_3$]I$_3$ and monoclinic [RuPr$_3$]I$_3$ types of structures for systems with iron-group elements as endohedral atoms suggests that there is a competition to adopt the respective type of structure. An examination [72] of the electronic band structures for two isocompositional bromides of both $[TR_3]X_3$-types reveals that the Fermi level of the cubic compound falls in a maximum of the densities-of-states (DOS), while the Fermi level in the monoclinic representative resides in a gap. Because the location of the Fermi level at a peak of the DOS curves typically indicates an electronically unfavorable situation, the monoclinic [RuPr$_3$]I$_3$-type structure should be preferred. Why, then, do we observe a competition between the monoclinic [RuPr$_3$]I$_3$-type and the cubic [PtPr$_3$]I$_3$-type for early lanthanide cluster bromides and iodides comprising iron-group elements as interstitials?

Table 1. Average −ICOHP/bond values and percentage contributions of the respective interactions to the net bonding capabilities in rare-earth transition-metal halides composed of tetrameric clusters, $[T_4R_{16}]$, and cluster chains. The details about the quantum-chemical computations, distance ranges, −ICOHP/bond ranges, and cumulative −ICOHP/cell values may be extracted from the respective literature listed in the last column. [a] hypothetical monoclinic $[TR_3]X_3$-type halide; details regarding the generation of the model may be obtained from the respective literature.

Compound	$R-T$		$R-R$		$R-X$		$T-T$		Ref.
	Ave. −ICOHP/ Bond (eV/bond)	%	Ave. −ICOHP/ Bond (eV/bond)	%	Ave. −ICOHP/ Bond (eV/bond)	%	Ave. −ICOHP/ Bond (eV/bond)	%	
Rare-earth transition-metal halides comprising tetramers									
$[Ru_4Y_{16}]Br_{20}$	2.03	48.8	0.11	5.2	0.60	43.2	0.48	2.9	[66]
$[Ru_4Y_{16}]I_{20}$	1.89	50.1	0.08	4.5	0.56	44.2	0.18	1.2	[66]
$[Ir_4Y_{16}]Br_{24}$	2.03	41.9	0.08	3.5	0.80	51.8	0.60	2.8	[66]
$[Ru_4Ho_{16}]I_{24}(Ho_4I_4)$	2.02	39.0	0.09	4.1	0.71	54.9	0.42	2.0	[66]
$[Ir_4Tb_{16}]Cl_{24}(TbCl_3)_4$	2.12	35.3	0.10	3.4	0.87	57.2	0.86	3.6	[66]
$[Rh_4Tb_{16}]Br_{24}(TbBr_3)_4$	2.17	34.0	0.14	4.5	0.95	59.6	0.49	1.9	[66]
$[Ir_4Tb_{16}]Br_{24}(TbBr_3)_4$	2.41	37.4	0.11	3.7	0.90	56.0	0.77	3.0	[66]
$[Ir_4Sc_{16}]Cl_{24}(ScCl_3)_4$	2.16	33.6	0.08	2.7	0.95	59.1	1.25	4.9	[66]
$[Os_4Sc_{16}]Cl_{24}(ScCl_3)_4$	2.26	33.5	0.09	2.7	0.98	58.0	1.57	5.8	[66]
$[Ru_4Sc_{16}]Cl_{24}(ScCl_3)_4$	2.08	31.6	0.10	3.3	1.01	61.1	1.07	4.0	[66]
$[Ru_4Gd_{16}]Br_{24}(GdBr_3)_4$	2.47	35.5	0.16	4.8	1.00	57.6	0.62	2.2	[66]
Rare-earth transition-metal halides comprising cluster chains									
c-[RuLa$_3$]Br$_3$	1.91	65.0	0.07	4.2	0.45	30.7	0.003	0.1	[72]
m-[RuLa$_3$]Br$_3$ [a]	2.37	64.3	0.18	9.1	0.49	26.5	−0.01	0.1	[72]
c-[IrLa$_3$]Br$_3$	1.87	67.2	0.06	3.6	0.40	29.1	0.004	0.1	[72]
m-[RuLa$_3$]I$_3$	2.45	60.8	0.22	9.9	0.59	29.2	−0.003	0.1	[72]
o-[RuPr$_3$]Cl$_3$	1.30	52.5	0.11	4.3	0.37	30.0	1.15	13.2	[73]
m-[RuPr$_3$]I$_3$	2.14	62.7	0.20	8.7	0.48	28.4	0.03	0.26	[73]
m-[MnGd$_3$]I$_3$	1.45	39.4	0.39	10.5	0.73	34.1	2.06	16.0	[73]

A chemical bonding analysis based on the COHP curves (Figure 2) and their integrated values (ICOHP; Table 1) for both compounds reveals that the heteroatomic $R-T$ and $R-X$ interactions show the largest percentages to the net bonding capabilities of the respective halides. A topological inspection of the respective local atomic environments indicates that the monoclinic representative contains less 'polar' heteroatomic contributions, which are maximized in the cubic structure. Accordingly, the interplay between the attempts to accomplish an electronically favorable situation and to optimize overall bonding appears to regulate the structural preferences between the cubic and monoclinic structures. Further research on the electronic structure of a cubic [PtPr$_3$]I$_3$-type bromide containing a cobalt-group element as endohedral atom bared that the Fermi level falls in a pseudogap, indicating an electronically favorable situation. Notably, the structures of the [RuPr$_3$]Cl$_3$ and [IrY$_3$]I$_3$ types, which are observed for the chlorides of the light lanthanides and the bromides as well as iodides of the heavier lanthanides, respectively [71,74,75], can be derived from that of the monoclinic [RuPr$_3$]I$_3$-type through displacements of the metal chains such that double chains of transition-metal centered rare-earth unicapped trigonal prims evolve. An examination [73] of the COHP curves and ICOHP values for representatives of the [RuPr$_3$]Cl$_3$ and [IrY$_3$]I$_3$ types of structure (Table 1) brings to light that significant bonding interactions are evident for the $T-T$ contacts within the cluster chains showing noticeable contributions to the net bonding capabilities; yet such percentages to net bonding capabilities are not identified for the large $T-T$ separations in the cluster chains of the cubic [PtPr$_3$]I$_3$ and monoclinic [RuPr$_3$]I$_3$ types.

In summary, the outcome of the bonding analyses employing the COHP method for the examples of the most prolific representatives of the transition-metal centered rare-earth cluster halides indicates the dominant role of the heteroatomic $R-T$ and $R-X$ interactions in these materials. Under consideration of the topologies and polarities in *Werner*-type complexes, the results of the bonding analyses imply that the transition-metal centered rare-earth clusters can also be described as anti-*Werner* complexes, in which the central atoms have higher electronegativities than the surrounding ligands [54].

3.2. The Bonding Situations in Electron-Poorer Polar Intermetallics Containing Gold

As signaled in the previous section for the examples of the isolated transition-metal cluster halides, there have been different approaches to develop effective electron counting rules, which were considered to help recognizing bonds in solid state materials [76]. For instance, *Wade's* rules [77,78] or the *Zintl* concept [5–10] are prominent electron-counting schemes being typically applied to recognize the valence bonds in solid-state materials with polyanionic clusters and monoatomic counterions. More recent research on the components of the active-metal (main-groups I, II, and the scandium-group elements)−gold−post-transition-metal systems identified several materials composed of polyanionic clusters possessing fewer valence electrons relative to those in *Zintl* phases [15]. In fact, the valence-electron concentrations of these materials called electron-poorer intermetallics place them close to the *Hume–Rothery* phases [11–13]. (Figure 3; a list of e/a ratios is provided in Table 2) [79] The propensity of gold to be incorporated in these polyanionic clusters and to contribute in hetero- as well as homoatomic bonding is expected to stem from the impact of relativistic effects [80–83] leading to a tighter binding of the 6s orbitals to the Au atom, a less tight binding of the 5d orbitals, and, ultimately, more 6s−5d orbital mixing. In a lack of electron counting rules, which help to recognize the bonds in these materials, the information of the locations and types of bonds in the electron-poorer intermetallics need to be gained from the computed electronic band structures. In this section, we present a number of examples, in which the locations and types of bonds were revealed with the aid of the COHP method.

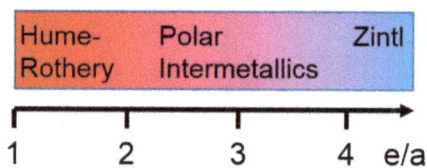

Figure 3. Positions of the *Zintl* phases, polar intermetallics, and *Hume–Rothery* phases with respect to their respective valence-electron concentrations (electrons per atom = e/a) [45,84].

Based on the topologies of the polyanionic clusters, each of the ternary compounds composed of an active-metal, gold, and a post-transition-element can be assigned to one of the three following groups: (I) compounds composed of one-dimensional polyanionic tunnels; (II) intermetallics constructed of hexagonal-diamond-fashioned polyanionic networks; and (III) compounds comprising polyanionic clusters in the forms of diverse (types of) polyhedra. Because the crystal structures of certain electron-poorer intermetallics presented in this survey possess atomic positions exhibiting occupational and/or positional disorders, the electronic band-structure calculations and bonding analyses were accomplished based on hypothetical models that approximate the actual crystal structures and usually show the lowest total energies. To develop a starting model for materials with disordered atomic sites, it is convenient to screen diverse feasible models for the scheme with the lowest total energy, because the structure model that shows the lowest total energy (and electronic as well as dynamic stability) among diverse possible models is considered to be the most preferable to approximate the experimentally determined model [85–87]. Detailed information regarding the structure determinations, the crystal structures, and generating the hypothetical models employed for the electronic band structure computations may be extracted from the respective references.

Table 2. Average–ICOHP/bond values (eV/bond) and percentage contributions to the net bonding capabilities of the homoatomic Au−Au and post-transition-metal−post-transition-metal as well as heteroatomic Au−post-transition-metal interactions for diverse active-metal-poor polar intermetallics consisting of an active-metal (main-groups I, II, and scandium-group elements), gold and a post-transition-metal. The valence-electron concentrations (e/a) are given in the second column, while details regarding the quantum-chemical calculations and the crystal structures of the respective compounds may be extracted from the references listed in the last column [45].

Compound	e/a	Parent Compound Disordered?	Homoatomic Contacts		Heteroatomic Contacts		Ref.
			Ave. −ICOHP/Bond	%	Ave. −ICOHP/Bond	%	
Compounds with anionic fragments in the forms of 1D tunnels in the crystal structures							
$EuAu_5In$	1.43	y/$EuAu_{5.0}In_{1.0}$	Au−Au: 0.79	57.1	Au−In: 0.81	36.5	[88]
KAu_3Ga_2	1.67	y/$KAu_{3.1}Ga_{1.9}$	Au−Au: 0.79 / Ga−Ga: 0.55	20.1 / 5.6	Au−Ga: 1.18	72.2	[89]
$RbAu_3Ga_2$	1.67	n	Au−Au: 0.66 / Ga−Ga: 0.53	17.7 / 5.7	Au−Ga: 1.17	75.4	[90]
$Na_{0.5}Au_2Ga_2$	1.89	y/$Na_{0.6}Au_2Ga_2$	Au−Au: 1.00 / Ga−Ga: 0.63	10.2 / 6.5	Au−Ga: 1.31	80.9	[90]
$K_{0.5}Au_2Ga_2$	1.89	y/$K_{0.6}Au_2Ga_2$	Au−Au: 0.97 / Ga−Ga: 0.51	8.5 / 4.4	Au−Ga: 1.64	85.7	[89]
$Rb_{0.5}Au_2Ga_2$	1.89	y/$Rb_{0.6}Au_2Ga_2$	Au−Au: 1.02 / Ga−Ga: 0.62	9.9 / 5.9	Au−Ga: 1.43	83.1	[90]
$NaAu_2Ga_4$	2.14	no	Ga−Ga: 1.04	20.0	Au−Ga: 1.73	72.2	[91]
KAu_2Ga_4	2.14	y/$KAu_{2.2}Ga_{3.8}$	Au−Au: 1.04 / Ga−Ga: 1.20	1.6 / 22.7	Au−Ga: 1.88	71.3	[89]
$CsAu_5Ga_9$	2.20	no	Au−Au: 0.59 / Ga−Ga: 0.48	2.7 / 16.2	Au−Ga: 1.42	78.8	[92]
Compds. with hexagonal diamond-type networks as anionic fragments in the crystal structures							
$Sr_2Au_7Zn_2$	1.36	y/$Sr_2Au_6(Au,Zn)_3$	Au−Au: 1.21 / Zn−Zn: 0.56	40.5 / 1.0	Au−Zn: 1.02	37.7	[93]
$Sr_2Au_7Al_2$	1.55	y/$Sr_2Au_{7.3}Al_{1.7}$	Au−Au: 1.07 / Al−Al: 1.26	32.0 / 2.4	Au−Al: 1.65	43.0	[94]
$SrAu_5Al_2$	1.63	y/$SrAu_{5.05}Al_{1.95}$	Au−Au: 1.01 / Al−Al: 0.71	32.0 / 1.6	Au−Al: 1.61	54.6	[94]
$Sr_2Au_6Al_3$	1.73	y/$Sr_2Au_{6.2}Al_{2.8}$	Au−Au: 1.09 / Al−Al: 1.56	21.5 / 7.7	Au−Al: 1.68	50.0	[94]
$SrAu_4Al_3$	1.88	y/$SrAu_{4.1}Al_{2.9}$	Au−Au: 0.93 / Al−Al: 1.48	17.0 / 8.9	Au−Al: 1.61	63.0	[94]
Compounds with diverse (types of) polyhedrons formed by the anions in the crystal structures							
$K_{12}Au_{21}Sn_4$	1.32	no	Au−Au: 1.22	28.0	Au−Sn: 2.70	43.1	[95]
$Na_8Au_{11}Ga_6$	1.48	y/$Na_8Au_{10.1}Ga_{6.9}$	Au−Au: 1.22 / Ga−Ga: 1.49	31.9 / 5.2	Au−Ga: 1.71	44.7	[96]
$NaAu_4Ga_2$	1.57	no	Au−Au: 1.20 / Ga−Ga: 0.51	27.9 / 1.2	Au−Ga: 1.61	65.0	[91]
Y_3Au_9Sb	1.77	no	Au−Au: 1.17	51.7	Au−Sb: 1.06	11.7	[97]
$CaAu_4Bi$	1.83	y/$CaAu_{4.1}Bi_{0.9}$	Au−Au: 1.40	57.5	Au−Bi: 0.54	22.2	[98]
$EuAu_6Al_6$	2.00	y/$EuAu_{6.1}Al_{5.9}$	Au−Au: 0.88 / Al−Al: 0.95	11.6 / 10.8	Au−Al: 1.58	67.8	[99]
$EuAu_6Ga_6$	2.00	y/$EuAu_{6.2}Ga_{5.8}$	Au−Au: 0.64 / Ga−Ga: 0.91	11.2 / 11.1	Au−Ga: 1.40	68.0	[99]
$Na_5Au_{10}Ga_{16}$	2.03	no	Au−Au: 0.42 / Ga−Ga: 1.10	0.8 / 22.2	Au−Ga: 1.67	71.2	[91]
$Y_3Au_7Sn_3$	2.15	no	Au−Au: 0.78	22.2	Au−Sn: 1.33	42.1	[100]
$Gd_3Au_7Sn_3$	2.15	no	Au−Au: 0.78	21.9	Au−Sn: 1.31	41.1	[100]

The crystal structures of the polar intermetallic compounds belonging to the first of the three categories are composed of cages which are assembled by the gold and post-transition-metal atoms and condensed along a particular crystallographic direction to yield one-dimensional tunnels. These cages enclose the active-metals that are stringed along, or slightly displaced from, the axes running through the one-dimensional tunnels parallel to the given crystallographic paths such that the cages and the encased active-metals may show certain shifts from perfect linear chains for some of the representatives. For instance, the crystal structure of $Sr_3Au_8Sn_3$ [101] contains one-dimensional tunnels of pentagonal and hexagonal prisms that are assembled by the gold and tin atoms and surround the strontium atoms (Figure 4). A chemical-bonding analysis based on the COHP curves for this material indicates that a structural transformation from a high-temperature to a low-temperature polymorph of this compound is influenced by the trend to optimize the Au−Au and Au−Sn bonding interactions. The gold-rich $K_{1.8}Au_6In_4$ [102] is another example of a polar intermetallic compound composed of one-dimensional tunnels which are constructed by the gold and post-transition-metal atoms (Figure 4). An examination based on the ICOHP values for this compound denoted considerable Au−In bonding interactions besides weaker, but evident, homoatomic interactions. Notably, a survey of the average ICOHP/bond values and their percentages to the net bonding capabilities for the Au/post-transition-metal−Au/post-transition-metal interactions in polar intermetallics with one-dimensional polyanionic tunnels in their crystal structures brings to light that the largest ICOHP/bond values and percentage contributions typically stem for the heteroatomic contacts (Table 2).

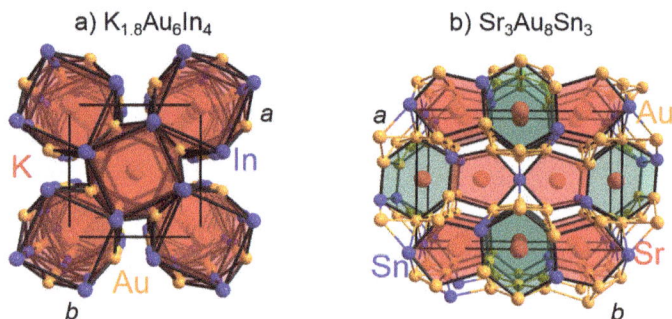

Figure 4. Representations of the crystal structures of (**a**) $K_{1.8}Au_6In_4$ and (**b**) $Sr_3Au_8Sn_3$, in which the gold (yellow) and post-transition-metal atoms (blue) assemble tunnels encompassing the active-metal atoms (red).

The second of the three aforementioned classes of polar intermetallic compounds contains those materials composed of hexagonal diamond-like gold networks, which have so far been identified to be present in the crystal structures of four different types of polar intermetallics with diverse combinations of post-transition-elements (Zn, Cd, Al, Ga, In, or Sn) and active-metals (Sr, Ba, Eu) [93,94,103–107]. The cavities within the hexagonal-diamond-like gold networks encompass the active-metal atoms, or triangles assembled by extra gold and post-transition-metal atoms (Figure 5). The COHP curves and their respective integrated values have been examined for diverse types of polar intermetallics containing hexagonal-diamond-type gold networks (see Table 2 for reported ICOHP/bond values and the respective percentages). For instance, an investigation [107] of the COHP curves for $BaAu_5Ga_2$ and $BaAu_4Ga_3$ bared that the broad majority of the bonding interactions resides between the Au−Au and Au−Ga interactions. In that connection, it is remarkable that the contributions of the homoatomic and heteroatomic contacts within the hexagonal-diamond-type networks and triangles are comparable for some of these compounds, while the largest percentages to the total bonding capabilities in the electron-poorer intermetallics often stem from the heteroatomic gold−post-transition-metal interactions (see Table 2).

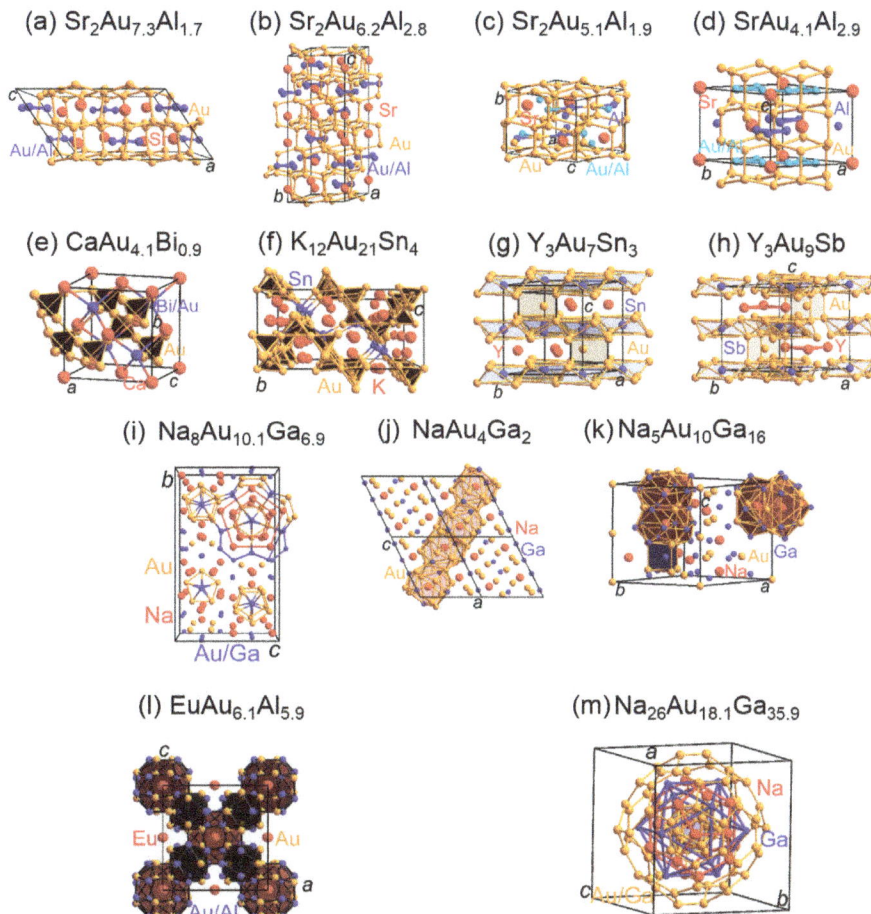

Figure 5. Representations of the crystal structures of (**a**) $Sr_2Au_{7.3}Al_{1.7}$; (**b**) $Sr_2Au_{6.2}Al_{2.8}$; (**c**) $Sr_2Au_{5.1}Al_{1.9}$; (**d**) $SrAu_{4.1}Al_{2.9}$; (**e**) $CaAu_{4.1}Bi_{0.9}$; (**f**) $K_{12}Au_{21}Sn_4$; (**g**) $Y_3Au_7Sn_3$; (**h**) Y_3Au_9Sb; (**i**) $Na_8Au_{10.1}Ga_{6.9}$; (**j**) $NaAu_4Ga_2$; (**k**) $Na_5Au_{10}Ga_{16}$; (**l**) $EuAu_{6.1}Al_{5.9}$; and (**m**) $Na_{26}Au_{18.1}Ga_{35.9}$. In the case of $Na_{26}Au_{18.1}Ga_{35.9}$, the diverse cluster shells typically observed for Bergman-type quasicrystals are shown, while atoms which are located in the unit cell but do not assemble the cluster shells have been omitted for the benefit of a clear representation. Details regarding the crystal structures and their determinations may by extracted from the literature cited in the main text and in the Table 2. Parts of the figure are adapted from reference [45].

The third class of three aforementioned groups of electron-poorer polar intermetallics contains those compounds whose crystal structures feature polyhedral clusters of the anionic components surrounding the active-metal and/or extra gold or post-transition-metal atoms. In particular, these compounds cannot be assigned to one of the other groups based on the spatial arrangements of the polyanionic clusters in the crystal structures of these polar intermetallics. Because of the complexity of the crystal structures showing defects such as disorders of vacancies, some components of this group can also be assigned to the broad family of complex metallic alloys [108,109]. For instance, the crystal structures of $EuAu_{6.1}Al_{5.9}$ and $EuAu_{6.2}Ga_{5.8}$ are both derived from the $NaZn_{13}$-type and include icosahedra, tetrahedral stars, and europium-centered snub cubes formed by the gold and post-transition-metal atoms (Figure 5) [99]. An examination of the COHP curves and ICOHP values for models approximating the real crystal structures of these intermetallics demonstrates that the driving force stabilizing these materials stems from the maximization of the amounts of the heteroatomic Au−post-transition-metal contacts. More recent research on the sodium−gold−gallium system identified an icosahedral (Bergman-type) quasicrystal, i.e., $Na_{13}Au_{12}Ga_{15}$, for which a COHP bonding analysis revealed extensive Au/post-transition-metal−Au/post-transition-metal bonding and, also, contributions from sodium in delocalized metal−metal bonding [84]. Additional investigations of the sodium−gold−gallium system resulted in the discoveries of a series of intermetallics, i.e., $Na_8Au_{10.1}Ga_{6.9}$, $NaAu_4Ga_2$, and $Na_5Au_{10}Ga_{16}$, composed of diverse sorts of polyhedra that are constituted by the gold and gallium atoms and enclose the sodium atoms [91,96]. The chemical bonding analyses were accomplished by means of the COHP method for these intermetallic compounds and exhibited extensive gold/post-transition-metal−gold/post-transition-metal bonding for these materials. $CaAu_{4.1}Bi_{0.9}$ is an example of a polar intermetallic compound, in which the largest percentages of the cumulative ICOHP per cell to the total bonding capabilities originate from the Au−Au interactions [98]. The crystal structure of the bismuth-containing intermetallic is derived from the $MgCu_2$-type and constructed of networks of vertices-sharing gold tetrahedra enclosing the calcium and bismuth atoms. Notably, the structural motif of vertices-sharing gold tetrahedra has also been encountered for other types of polar intermetallics as, e.g., $K_{12}Au_{21}Sn_4$ [95,110] Furthermore, gold clusters have been identified for the crystal structures of $R_3Au_7Sn_3$ (R = Y, Gd) and Y_3Au_9Sb, in which the gold atoms assemble trigonal prisms and antiprisms enclosing extra gold and post-transition-metal atoms, respectively [97,100]. In the antimony-containing compound, the largest percentage contributions of the cumulative ICOHP per cell to the net bonding capabilities arise from the Au−Au interactions, while the largest shares of the cumulative ICOHP per cell to the total bonding capabilities in the tin-containing compounds stem from the Au−Sn separations (Table 2).

In summary, the COHP method has been demonstrated to be a beneficial means for identifying the bonding situations in electron-poorer, polar intermetallic compounds composed of gold. Under consideration of the outcome of the bonding analyses reported for these materials to date, the largest ICOHP per cell are often observed for the heteroatomic gold−post-transition-metal contacts showing the largest percentages to the net bonding capabilities in these compounds.

3.3. The Role of Vacancies and Structural Preferences in Phase-Change Materials

The group of phase-change materials comprises those "intermetallic" compounds which can reversibly transform from amorphous to crystalline phase after irradiation with laser light [111–114]. Because the crystalline and amorphous phases of phase-change materials significantly differ in both reflectivity and resistance, these compounds can be utilized as rewriteable data-storage materials [111–113]. In particular, an application of a long pulse of a low-intensity laser beam heats the amorphous regions of the materials leading to recrystallizations (set pulse), while an application of a short pulse of a high-intensity laser beam locally melts the crystalline material forming amorphous regions after fast quenching (reset pulse) [111]. After having identified fast recrystallization and good optical contrast for GeTe and $Ge_{11}Te_{60}Sn_4Au_{25}$ [115,116], diverse phase-change materials composed of different combinations of a tetrel element, a pnictogen, and tellurium have been discovered to date [117]. The quest for previously unknown materials suited for phase-change memory applications stimulated the impetus to identify the origins of the reversible phase transitions typically observed for phase-change materials. Because the measurements of the dielectric functions for phase-change materials revealed that the optical dielectric constants are evidently higher for crystalline than for the amorphous phases, it was inferred that the sorts of bonding significantly change between the two phases [118]. The subsequent developments of materials´ maps which were based on the ionicities and hybridizations determined for the bonds in diverse compounds indicated that the phase-change materials are evident solely for small segments in these maps [117,119]. Additional examinations on the nature of bonding for the local structural arrangements in the amorphous phase of the phase-change material GeTe revealed that the tetrahedral fragments in the crystal structure are stabilized by the homoatomic Ge−Ge bonds [120]. In this section, we will present two applications of the COHP method to disclose the structural preferences for phase-change materials.

A prototypical representative of the group of phase-change materials is the ternary $Ge_2Sb_2Te_5$, which crystallizes with a rocksalt-type of structure in its metastable state [121]. Notably, an exanimation of that crystal structure for $Ge_2Sb_2Te_5$ revealed that vacancies of about 20 at.-% are evident for the occupationally disordered Ge/Sb sites [122]. To understand the origin of the presence of vacancies for the mixed Ge/Sb sites in the rocksalt-like structure of $Ge_2Sb_2Te_5$, the bonding situations were examined based on the COHP curves of three different compositions, i.e., $Ge_2Sb_2Te_4$, $Ge_{1.5}Sb_2Te_4$, and $GeSb_2Te_4$ [121] A comparison of the COHP curves for the three different compositions (Figure 6) reveals that significantly antibonding Ge−Te and Sb−Te interactions are evident at the Fermi level in the germanium-richest telluride, while less and no antibonding Ge−Te and Sb−Te states are present at the Fermi levels in the germanium-poorer $Ge_{1.5}Sb_2Te_4$ and $GeSb_2Te_4$, respectively. Accordingly, it can be inferred that the presence of vacancies in the structure of the ternary phase-change materials corresponds to a reduction of the valence-electron concentration in order to optimize the overall bonding. In other words, nature gets rid of the antibonding states by expelling some of the electron-donating Ge atoms.

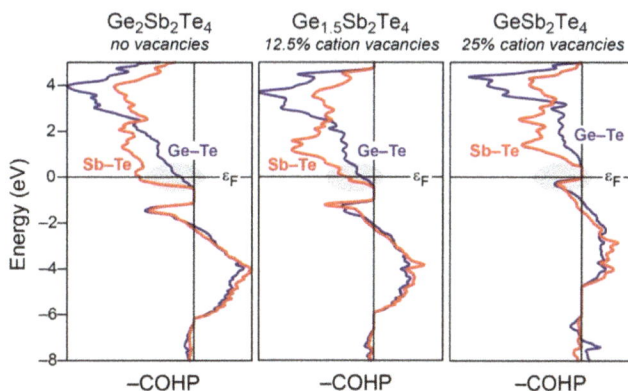

Figure 6. COHP curves of $Ge_2Sb_2Te_4$ (**left**); $Ge_{1.5}Sb_2Te_4$ (**middle**); and $GeSb_2Te_4$ (**right**) [121]: the Ge−Te and Sb−Te COHP curves are shown in blue and red, respectively, while the Fermi levels are represented by the black horizontal lines.

Recent research in the field of phase-change data storage materials has focused on the developments of materials suited for applications in storage-class memories beyond the scope of traditional phase-change materials. In that connection, chalcogenide superlattices (CSL) [123,124] and nanocrystals [125,126] were investigated for their capabilities to serve in such systems. In particular, interfacial phase-change materials (iPCMs) [127–129], in which the transformations do not occur between an amorphous and a crystalline phase, but between two crystalline phases, are of great interest as candidate systems. The more recent determinations of the crystal structure for the ternary Ge_4Se_3Te [31] (Figure 7) indicated that the type of structure observed for that chalcogenide may serve as an archetype for future PCMs. In particular, the crystal structure of Ge_4Se_3Te is composed of layers of germanium and the chalcogenides (Ch = Se/Te; note that the chalcogenide sites are occupationally disordered) with the stacking sequence of −Ch−Ge−Ge−Ch−. The stacking sequence of the germanium and chalcogenide layers in Ge_4Se_3Te is in stark contrast to that observed for α-GeTe, in which the layers of the tellurium and germanium atoms are arranged in a stacking sequence of $-Ge-Te-Ge-Te-$. At this point, one may wonder why there is a difference between the stacking sequences of the germanium and chalcogenide layers in Ge_4Se_3Te and α-GeTe.

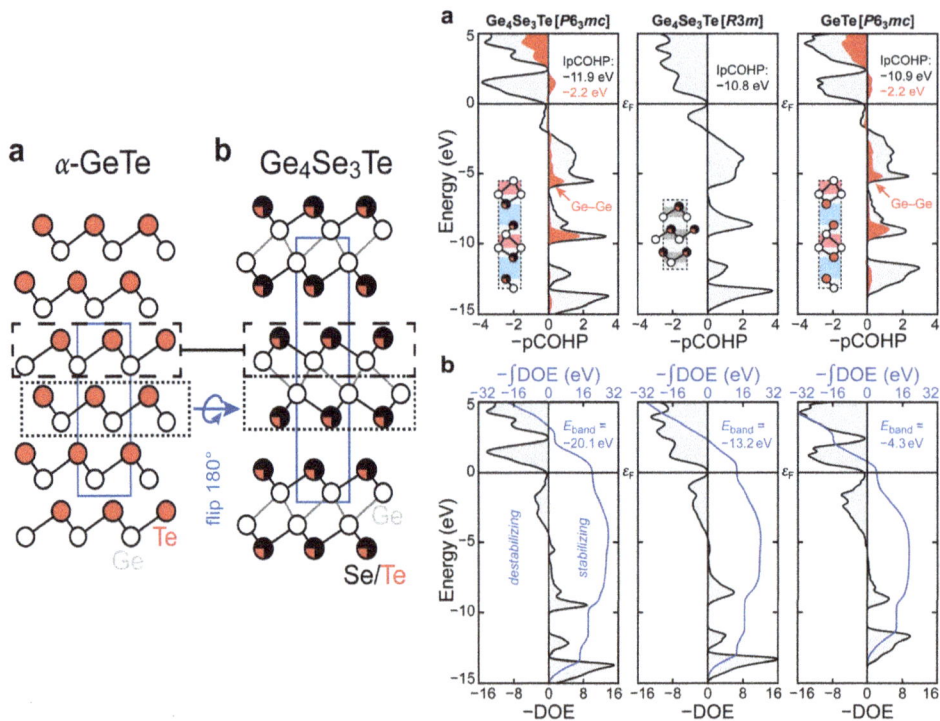

Figure 7. (**Left**): Representations of the crystal structures of and structural relationships between (**a**) α-GeTe and (**b**) Ge_4Se_3Te: the crystal structure of the two chalcogenides discern in the stacking sequences of the layers of the germanium atoms and the chalcogenide atoms. (**Right**): (**a**) pCOHP curves and (**b**) DOE functions of Ge_4Se_3Te, a α-GeTe-type Ge_4Se_3Te, and a Ge_4Se_3Te-type GeTe: the Fermi levels, E_F, are represented by the black horizontal lines. Reprinted with permission from reference [31]. Copyright 2017, WILEY-VCH, Weinheim.

To understand the origin of the differences between the stacking sequences of the germanium and chalcogenide layers in Ge_4Se_3Te and α-GeTe, the COHP curves and Density-of-Energy (DOE) functions of Ge_4Se_3Te, an α-GeTe-type Ge_4Se_3Te, and a Ge_4Se_3Te-type GeTe were examined. The COHP curves for Ge_4Se_3Te and the α-GeTe-type Ge_4Se_3Te (Figure 7) indicate that there are antibonding Ge$-$Ch interactions at the Fermi level in the α-GeTe-type chalcogenide, while the antibonding Ge$-$Ch interactions are weaker at the Fermi level in the Ge_4Se_3Te crystallizing with its own type of structure, because the electrons have been reshuffled into Ge$-$Ge bonding interactions. Thus, the electronically unfavorable situation corresponding to antibonding states at the Fermi level is not alleviated by introducing vacancies as demonstrated in the previous paragraph, but by transferring electrons in tetrel$-$tetrel bonding states in a different type of structure. From a comparison of the DOE functions and their integrated values, it is even clearer that the adoption of its own type of structure instead of the α-GeTe-type is favored for Ge_4Se_3Te because of a significant loss of destabilizing interactions from the α-GeTe-type to the Ge_4Se_3Te-type. An inspection of the DOE function for a Ge_4Se_3Te-type

GeTe reveals a lack of stabilizing energy contributions at lower energies and strongly destabilizing contributions at the Fermi level such that the Ge_4Se_3Te-type is not preferred for GeTe.

3.4. Itinerant Antiferromagnetism and Ferromagnetism from the Viewpoint of Chemical Bonding in Intermetallic Compounds

One of the very first approaches to conveniently describe and account for the occurrence of ferromagnetic states in metals and their compounds was provided by the *chemical* theory of the collective electron ferromagnetism, which establishes a relationship between the incidence of ferromagnetism and the electronic band structures in metals and their compounds [130]. At this point, one may wonder how the presence of itinerant antiferromagnetism and ferromagnetism in metals and their compounds may be viewed from the perspective of chemical bonding. Indeed, the COHP technique can also be applied to the outcome of spin-polarized computations, thereby providing valuable information regarding the relationships between the itinerant magnetic states and the bonding situation in a given solid-state material. In this section, we will present some applications of the COHP method to spin-polarized first-principles-based computations on intermetallics.

An inspection of the non-spin-polarized COHP curves for $3d$ transition-metals (T; Figure 8) indicates that the Fermi levels in the ferromagnetic metals cross antibonding regions of the $T-T$ COHP curves [131,132]; however, in the case of the spin-polarized COHP curves for the ferromagnetic metals, the states occupied by the majority α spins lower in energy, while the states comprising the minority β spins shift upward in energy. As a result, the formerly antibonding states at the Fermi levels have disappeared and the $T-T$ bonding has been strengthened (e.g., by 5% in bcc-Fe). The strengthening of the chemical bonding upon onset of ferromagnetism is smaller for cobalt and even smaller for nickel, and their magnetic moments and exchange splitting are also smaller. On the other hand, positioning the Fermi level in the nonbonding levels (such as for bcc-Cr) is a clear signpost for antiferromagnetism. Hence, chemically fine-tuning the valence-electron concentration (and, because of that, the Fermi level) to properly "hit" antibonding and nonbonding states yields a simple but rational recipe for making ferromagnets and antiferromagnets [133], even entire series of such materials. For instance, chemical bonding analyses using the COHP technique provided fruitful insights into the origins of the magnetic ground states for several (intermetallic) borides [134–137] such that the magnetic orderings of the respective borides may be "tailored".

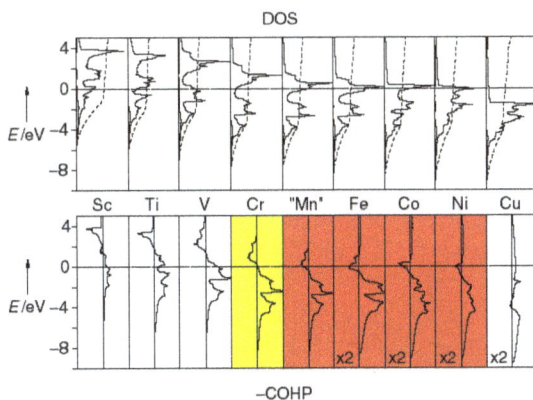

Figure 8. DOS and transition-metal–transition-metal COHP curves of the $3d$ transition-metals in a non-magnetic regime: the COHP curves shaded in red correspond to transition-metals showing paramagnetic-ferromagnetic transitions, while the COHP curve of chromium exhibiting a paramagnetic-antiferromagnetic transition is shaded in yellow. Reprinted with permission from reference [131]. Copyright 2000 WILEY-VCH, Weinheim.

Among the diverse binary compounds consisting of one of the 3*d* transition-metal, the nitrides are of fundamental importance for research as well as technical applications due to their roles in the production and hardening of steel [138]. In the case of the mononitrides, two different types of structure, i.e., the rock salt and zinc blende type of structures, have been proposed. To identify the structural preferences for the mononitrides, *T*N (*T* = Sc–Ni), the COHP curves of the nitrides were examined. A comparison of the COHP curves for the transition-metal mononitrides demonstrates that more antibonding *T*−*T* interactions are occupied as the valence-electron counts of the transition-metals increase. Furthermore, FeN and CoN tend to crystallize in the zinc blende-type of structure, because the antibonding *T*−*T* interactions are less pronounced in the zinc blende-type than in the rock salt-type of structure.

More recent research on the iron nitrides identified a previously unknown NiAs-type modification for FeN, which was obtained from high-temperature high-pressure syntheses [139]. An inspection of the non-spin-polarized DOS and projected COHP curves for the NiAs-type FeN (Figure 9) indicates an electronically unfavorable situation since the Fermi level falls in a maximum of the DOS and significantly antibonding Fe−N interactions. Because the Fermi level falls in minima of the spin-polarized DOS curves and the integrated values of the spin-polarized pCOHP denote net bonding characters for the Fe−N interactions, it can be inferred that the NiAs-type FeN alleviates the electronically unfavorable situation by approaching a magnetic state, once again. This conclusion is corroborated by the presence of a magnetic sextet in the Mößbauer spectrum of the NiAs-type FeN. An additional comparison of the integrated COHP values in the NiAs-type FeN to those in the zinc blende-type modification shows that the zinc blende-type FeN comprises shorter Fe−N contacts providing larger integrated pCOHP values relative to the NiAs-type compound and, hence, should be preferred.

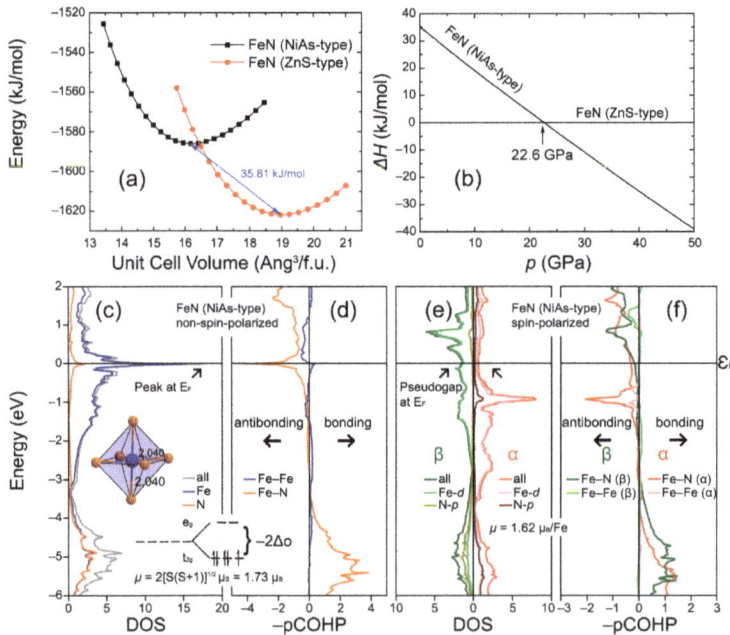

Figure 9. Energy-volume curves (**a**), and pressure-dependence of the enthalpies (**b**) of the NiAs-type and ZnS-type polymorphs of FeN; non-spin-polarized and spin-polarized densities-of-states (**c**,**e**), and pCOHP curves (**d**,**f**) of the NiAs-type FeN: the Fermi level, E_F, is represented by the black horizontal line. Reprinted with permission from reference [139]. Copyright 2017 WILEY-VCH, Weinheim.

4. Conclusions

Because the information concerning the spatial arrangements of elements in a given material is of fundamental interest for the materials design, there is critical need to recognize the bonding situation in a given material in order to understand the structural arrangements and preferences. The extraction of the information regarding the chemical bonding from electronic-structure computations requires the use of efficient and reliable procedures. Since its introduction 25 years ago, the COHP method has been employed to identify the bonding situations in numerous solid-state compounds (and also molecules). As shown in this contribution, the COHP technique does not depend on the employed basis set and can be applied to the results of electronic-structure computations obtained using diverse quantum-chemical means. Furthermore, we demonstrated the applications of this method to identify the bonding situations in diverse (polar) intermetallic compounds, which are traditionally considered as black sheep in the light of valence-electron counting rules [76]. The COHP analyses of the rare-earth transition-metal halides enabled identification of these compounds as anti-*Werner*-fashioned complexes, whereas the applications of the COHP procedure to the electron-poorer polar intermetallic compounds indicated that these materials tend to optimize the overall bonding by maximizing the amounts of heteroatomic bonds providing the largest ICOHP per bond values. Furthermore, chemical-bonding analyses based on the COHP technique have allowed identifying the structural preferences for nitrides and phase-change materials that are relevant for technical applications.

Author Contributions: This paper was completed trough contributions from both authors.

Acknowledgments: S.T. is grateful for a Liebig Stipend of the Verband der Chemischen Industrie e.V. (FCI), Frankfurt a. M.

Conflicts of Interest: The authors declare no conflict of interest.

References

1. Goethe, J.W. *Faust—Der Tragödie Erster Teil*; Reclam Publishers: Stuttgart, Germany, 1957.
2. Pauling, L. The Modern Theory of Valency. *J. Chem. Soc.* **1948**, 1461–1467. [CrossRef]
3. Lewis, G.N. The atom and the molecule. *J. Am. Chem. Soc.* **1916**, *38*, 762–785. [CrossRef]
4. Pauling, L. The principles determining the structure of complex ionic crystals. *J. Am. Chem. Soc.* **1929**, *51*, 1010–1026. [CrossRef]
5. Zintl, E. Intermetallische Verbindungen. *Angew. Chem.* **1939**, *52*, 1–6. [CrossRef]
6. Klemm, W.; Busmann, E. Volumeninkremente und Radien einiger einfach negativ geladener Ionen. *Z. Anorg. Allg. Chem.* **1963**, *319*, 297–311. [CrossRef]
7. Papoian, G.A.; Hoffmann, R. Hypervalent Bonding in One, Two, and Three Dimensions: Extending the Zintl-Klemm Concept to Nonclassical Electron-Rich Networks. *Angew. Chem. Int. Ed.* **2000**, *39*, 2408–2448. [CrossRef]
8. Köhler, J.; Whangbo, M.-H. Late transition metal anions acting as p-metal elements. *Solid State Sci.* **2008**, *10*, 444–449. [CrossRef]
9. Miller, G.J.; Schmidt, M.W.; Wang, F.; You, T.-S. Quantitative Advances in the Zintl-Klemm Formalism. In *Structure and Bonding*; Fässler, T., Ed.; Springer: Berlin/Heidelberg, Germany, 2011; Volume 139, pp. 1–55, ISBN 978-3-642-21150-8.
10. Nesper, R. The Zintl-Klemm Concept—A Historical Survey. *Z. Anorg. Allg. Chem.* **2014**, *640*, 2639–2648. [CrossRef]
11. Jones, H. The phase boundaries in binary alloys, Part 2: The theory of the α, β phase boundaries. *Proc. Phys. Soc.* **1937**, *49*, 250–257. [CrossRef]
12. Massalski, T.B.; Mizutani, U. Electronic Structure of Hume-Rothery Phases. *Prog. Mater. Sci.* **1978**, *22*, 151–262. [CrossRef]
13. Mizutani, U.; Sato, H. The Physics of the Hume-Rothery Electron Concentration Rule. *Crystals* **2017**, *7*, 9. [CrossRef]
14. Corbett, J.D. Exploratory Synthesis in the Solid State. Endless Wonders. *Inorg. Chem.* **2000**, *39*, 5178–5191. [CrossRef] [PubMed]

15. Corbett, J.D. Exploratory Synthesis: The Fascinating and Diverse Chemistry of Polar Intermetallic Phases. *Inorg. Chem.* **2010**, *49*, 13–28. [CrossRef] [PubMed]

16. Schrödinger, E. Quantisierung als Eigenwertproblem. *Ann. Phys.* **1926**, *384*, 361–376. [CrossRef]

17. Bloch, F. Über die Quantenmechanik der Elektronen in Kristallgittern. *Z. Physik* **1929**, *52*, 555–600. [CrossRef]

18. Dronskowski, R. *Computational Chemistry of Solid State Materials*; WILEY-VCH Publishers: Weinheim, Germany, 2005; ISBN 978-3-527-31410-2.

19. Hohenberg, P.; Kohn, W. Inhomogeneous Electron Gas. *Phys. Rev.* **1964**, *136*, B864–B871. [CrossRef]

20. Kohn, W.; Sham, L.J. Self-Consistent Equations Including Exchange and Correlation Effects. *Phys. Rev.* **1965**, *140*, A1133–A1138. [CrossRef]

21. Jones, R.O.; Gunnarsson, O. The density functional formalism, its applications and prospects. *Rev. Mod. Phys.* **1989**, *61*, 689–746. [CrossRef]

22. Dronskowski, R.; Blöchl, P.E. Crystal Orbital Hamilton Populations (COHP). *Energy-Resolved Visualization of Chemical Bonding in Solids Based on Density-Functional Calculations J. Phys. Chem.* **1993**, *97*, 8617–8624. [CrossRef]

23. Deringer, V.L.; Tchougréeff, A.L.; Dronskowski, R. Crystal Orbital Hamilton Population (COHP) Analysis As Projected from Plane-Wave Basis Sets. *J. Phys. Chem. A* **2011**, *115*, 5461–5466. [CrossRef] [PubMed]

24. Roothaan, C.C.J. New Developments in Molecular Orbital Theory. *Rev. Mod. Phys.* **1951**, *23*, 69–89. [CrossRef]

25. Hughbanks, T.; Hoffmann, R. Chains of Trans-Edge-Sharing Molybdenum Octahedra: Metal-Metal Bonding in Extended Systems. *J. Am. Chem. Soc.* **1983**, *105*, 3528–3537. [CrossRef]

26. Hoffmann, R. How Chemistry and Physics Meet in the Solid State. *Angew. Chem. Int. Ed. Engl.* **1987**, *26*, 846–878. [CrossRef]

27. Wolfsberg, M.; Helmholz, L. The Spectra and Electronic Structure of the Tetrahedral Ions MnO_4^-, CrO_4^-, and ClO_4^-. *J. Chem. Phys.* **1952**, *20*, 837–843. [CrossRef]

28. Slater, J.C. Wave Functions in a Periodic Potential. *Phys. Rev.* **1937**, *51*, 846–851. [CrossRef]

29. Andersen, O.K. Linear methods in band theory. *Phys. Rev. B Condens. Matter Mater. Phys.* **1975**, *12*, 3060–3083. [CrossRef]

30. Blöchl, P.E. Projector augmented-wave method. *Phys. Rev. B Condens. Matter Mater. Phys.* **1994**, *50*, 17953–17979. [CrossRef]

31. Küpers, M.; Konze, P.M.; Maintz, S.; Steinberg, S.; Mio, A.M.; Cojocaru-Mirédin, O.; Zhu, M.; Müller, M.; Luysberg, M.; Mayer, J.; et al. Unexpected Ge–Ge Contacts in the Two-Dimensional Ge_4Se_3Te Phase and Analysis of Their Chemical Cause with the Density of Energy (DOE) Function. *Angew. Chem. Int. Ed.* **2017**, *56*, 10204–10208. [CrossRef] [PubMed]

32. Kresse, G.; Marsman, M.; Furthmüller, J. *Vienna Ab Initio Simulation Package (VASP)*; The Guide Computational Materials Physics, Faculty of Physics, Universität Wien: Vienna, Austria, 2014.

33. Kresse, G.; Furthmüller, J. Efficiency of ab-initio total energy calculations for metals and semiconductors using a plane-wave basis set. *Comput. Mater. Sci.* **1996**, *6*, 15–50. [CrossRef]

34. Kresse, G.; Furthmüller, J. Efficient iterative schemes for ab initio total-energy calculations using a plane-wave basis set. *Phys. Rev. B Condens. Matter Mater. Phys.* **1996**, *54*, 11169–11186. [CrossRef]

35. Kresse, G.; Hafner, J. Ab initio molecular dynamics for liquid metals. *Phys. Rev. B Condens. Matter Mater. Phys.* **1993**, *47*, 558–561. [CrossRef]

36. Kresse, G.; Joubert, D. From ultrasoft pseudopotentials to the projector augmented-wave method. *Phys. Rev. B Condens. Matter Mater. Phys.* **1999**, *59*, 1758–1775. [CrossRef]

37. Gonze, X.; Jollet, F.; Abreu Araujo, F.; Adams, D.; Amadon, B.; Applencourt, T.; Audouze, C.; Beuken, J.-M.; Bieder, J.; Bokhanchuk, A.; et al. Recent developments in the ABINIT software package. *Comput. Phys. Commun.* **2016**, *205*, 106–131. [CrossRef]

38. Gonze, X.; Amadon, B.; Anglade, P.-M.; Beuken, J.-M.; Bottin, F.; Boulanger, P.; Bruneval, F.; Caliste, D.; Caracas, R.; Côté, M.; et al. ABINIT: First-principles approach to material and nanosystem properties. *Comput. Phys. Commun.* **2009**, *180*, 2582–2615. [CrossRef]

39. Gonze, X.; Rignanese, G.-M.; Verstraete, M.; Beuken, J.-M.; Pouillon, Y.; Caracas, R.; Jollet, F.; Torrent, M.; Zerah, G.; Mikami, M.; et al. A brief introduction to the ABINIT software package. *Z. Kristallog. Cryst. Mater.* **2005**, *220*, 558–562. [CrossRef]

40. Gonze, X.; Beuken, J.-M.; Caracas, R.; Detraux, F.; Fuchs, M.; Rignanese, G.-M.; Sindic, L.; Verstraete, M.; Zerah, G.; Jollet, F.; et al. First-principles computation of material properties: The ABINIT software projec. *Comput. Mater. Sci.* **2002**, *25*, 478–492. [CrossRef]

41. Giannozzi, P.; Baroni, S.; Bonini, N.; Calandra, M.; Car, R.; Cavazzoni, C.; Ceresoli, D.; Chiarotti, G.L.; Cococcioni, M.; Dabo, I.; et al. QUANTUM ESPRESSO: A modular and open-source software project for quantum simulations of materials. *J. Phys. Condens. Matter* **2009**, *21*, 395502. [CrossRef] [PubMed]

42. Giannozzi, P.; Andreussi, O.; Brumme, T.; Bunau, O.; Nardelli, M.B.; Calandra, M.; Car, R.; Cavazzoni, C.; Ceresoli, D.; Cococcioni, M.; et al. Advanced capabilities for materials modelling with Quantum ESPRESSO. *J. Phys. Condens. Matter* **2017**, *29*, 465901. [CrossRef] [PubMed]

43. Maintz, S.; Deringer, V.L.; Tchougréeff, A.L.; Dronskowski, R. Analytic Projection From Plane-Wave and PAW Wavefunctions and Application to Chemical-Bonding Analysis in Solids. *J. Comput. Chem.* **2013**, *34*, 2557–2567. [CrossRef] [PubMed]

44. Maintz, S.; Deringer, V.L.; Tchougréeff, A.L.; Dronskowski, R. LOBSTER: A Tool to Extract Chemical Bonding from Plane-Wave Based DFT. *J. Comput. Chem.* **2016**, *37*, 1030–1035. [CrossRef] [PubMed]

45. Gladisch, F.C.; Steinberg, S. Revealing Tendencies in the Electronic Structures of Polar Intermetallic Compounds. *Crystals* **2018**, *8*, 80. [CrossRef]

46. Schäfer, H.; Schnering, H.G. Metall-Metall-Bindungen bei niederen Halogeniden, Oxyden und Oxydhalogeniden schwerer Übergangsmetalle. *Angew. Chem.* **1964**, *76*, 833–849. [CrossRef]

47. Cotton, F.A. Strong Homonuclear Metal-Metal Bonds. *Acc. Chem. Res.* **1969**, *2*, 240–247. [CrossRef]

48. Simon, A. Cluster of Valence Electron Poor Metals—Structure, Bonding, and Properties. *Angew. Chem. Int. Ed. Engl.* **1988**, *27*, 159–183. [CrossRef]

49. Hughbanks, T. Bonding in clusters and condensed cluster compounds that extend in one, two and three dimension. *Prog. Solid St. Chem.* **1989**, *19*, 329–372. [CrossRef]

50. Corbett, J.D. Exploratory synthesis of reduced rare-earth-metal halides, chalcogenides, intermetallics: New compounds, structures, and properties. *J. Alloys Compds.* **2006**, *418*, 1–20. [CrossRef]

51. Corbett, J.D. Interstitially-stabilized cluster-based halides of the early transition metals. *J. Alloys Compds.* **1995**, *229*, 10–23. [CrossRef]

52. Simon, A.; Mattausch, H.J.; Ryazanov, M.; Kremer, R.K. Lanthanides as d Metals. *Z. Anorg. Allg. Chem.* **2006**, *632*, 919–929. [CrossRef]

53. Meyer, G. Reduced Halides of the Rare-Earth Elements. *Chem. Rev.* **1988**, *88*, 93–107. [CrossRef]

54. Meyer, G. Cluster Complexes as *anti-Werner* Complexes. *Z. Anorg. Allg. Chem.* **2008**, *634*, 2729–2736. [CrossRef]

55. Meyer, G. Rare Earth Metal Cluster Complexes. In *The Rare Earth Elements*; Atwood, D.A., Ed.; John Wiley & Sons, Ltd.: Chichester, UK, 2012; ISBN 978-1-119-95097-4.

56. Hughbanks, T.; Corbett, J.D. Encapsulation of Heavy Transition Metals in Iodide Clusters. Synthesis, Structure, and Bonding of the Unusual Cluster Phase $Y_6I_{10}Ru$. *Inorg. Chem.* **1989**, *28*, 631–635. [CrossRef]

57. Hughbanks, T.; Corbett, J.D. Rare-Earth-Metal Iodide Clusters Centered by Transition Metals: Synthesis, Structure, and Bonding of $R_7I_{12}M$ Compounds (R = Sc, Y, Pr, Gd; M = Mn, Fe, Co, Ni). *Inorg. Chem.* **1988**, *27*, 2022–2026. [CrossRef]

58. Rustige, C.; Brühmann, M.; Steinberg, S.; Meyer, E.; Daub, K.; Zimmermann, S.; Wolberg, M.; Mudring, A.-V.; Meyer, G. The Prolific $\{ZR_6\}X_{12}R$ and $\{ZR_6\}X_{10}$ Structure Types with Isolated Endohedrally Stabilized (Z) Rare-Earth Metal (R) Cluster Halide (X) Complexes. *Z. Anorg. Allg. Chem.* **2012**, *638*, 1922–1931. [CrossRef]

59. Hughbanks, T.; Rosenthal, G.; Corbett, J.D. Alloy Clusters: The Encapsulation of Transition Metals (Mn, Fe, Co, Ni) within Cluster Halides of Zirconium and the Rare-Earth Metals. *J. Am. Chem. Soc.* **1986**, *108*, 8289–8290. [CrossRef]

60. Sweet, L.E.; Roy, L.E.; Meng, F.; Hughbanks, T. Ferromagnetic Coupling in Hexanuclear Gadolinium Clusters. *J. Am. Chem. Soc.* **2006**, *128*, 10193–10201. [CrossRef] [PubMed]

61. Cotton, F.A.; Haas, T.E. A Molecular Orbital Treatment of the Bonding in Certain Metal Atom Clusters. *Inorg. Chem.* **1964**, *3*, 10–17. [CrossRef]

62. Ebihara, M.; Martin, J.D.; Corbett, J.D. Novel Chain and Oligomeric Condensed Cluster Phases for Gadolinium Iodides with Manganese Interstitials. *Inorg. Chem.* **1994**, *33*, 2079–2084. [CrossRef]

63. Payne, M.W.; Ebihara, M.; Corbett, J.D. A Novel Oligomer of Condensed Metal Atom Clusters in $[Y_{16}Ru_4I_{20}]$. *Angew. Chem. Int. Ed. Engl.* **1991**, *30*, 856–858. [CrossRef]

64. Steinwand, S.J.; Corbett, J.D. Oligomeric Rare-Earth-Metal Halide Clusters. Three Structures Built of $(Y_{16}Z_4)Br_{36}$ Units (Z = Ru, Ir). *Inorg. Chem.* **1996**, *35*, 7056–7067. [CrossRef] [PubMed]
65. Steinberg, S.; Zimmermann, S.; Brühmann, M.; Meyer, E.; Rustige, C.; Wolberg, M.; Daub, K.; Bell, T.; Meyer, G. Oligomeric rare-earth metal cluster complexes with endohedral transition metal atoms. *J. Solid State Chem.* **2014**, *219*, 159–167. [CrossRef]
66. Steinberg, S.; Bell, T.; Meyer, G. Electron Counting Rules and Electronic Structure in Tetrameric Transition-Metal (T)-Centered Rare-Earth (R) Cluster Complex Halides (X). *Inorg. Chem.* **2015**, *54*, 1026–1037. [CrossRef] [PubMed]
67. Dorhout, P.K.; Payne, M.W.; Corbett, J.D. Condensed Metal Cluster Iodides Centered by Noble Metals. Six Examples of Cubic R_3I_3Z Phases (R = La, Pr; Z = Os, Ir, Pt). *Inorg. Chem.* **1991**, *30*, 4960–4962. [CrossRef]
68. Zheng, C.; Mattausch, H.J.; Hoch, C.; Simon, A. La_3Br_3Ni: Jahn–Teller Distortion in the Reduced Rare Earth Metal Halide. *Z. Anorg. Allg. Chem.* **2009**, *635*, 2429–2433. [CrossRef]
69. Herzmann, N.; Mudring, A.-V.; Meyer, G. Seven-Coordinate Ruthenium Atoms Sequestered in Praseodymium Clusters in the Chloride {RuPr₃}Cl₃. *Inorg. Chem.* **2008**, *47*, 7954–7956. [CrossRef] [PubMed]
70. Payne, M.W.; Dorhout, P.K.; Kim, S.-J.; Hughbanks, T.R.; Corbett, J.D. Chains of Centered Metal Clusters with a Novel Range of Distortions: Pr_3I_3Ru, Y_3I_3Ru, and Y_3I_3Ir. *Inorg. Chem.* **1992**, *31*, 1389–1394. [CrossRef]
71. Steinberg, S.; Valldor, M.; Meyer, G. Change of magnetic and electronic features through subtle substitution in cubic, non-centrosymmetric extended rare-earth metal cluster complexes {TR₃}X₃. *J. Solid State Chem.* **2013**, *206*, 176–181. [CrossRef]
72. Steinberg, S.; Brgoch, J.; Miller, G.J.; Meyer, G. Identifying a Structural Preference in Reduced Rare-Earth Metal Halides by Combining Experimental and Computational Techniques. *Inorg. Chem.* **2012**, *51*, 11356–11364. [CrossRef] [PubMed]
73. Gupta, S.; Meyer, G.; Corbett, J.D. Contrasts in Structural and Bonding Representations among Polar Intermetallic Compounds. Strongly Differentiated Hamilton Populations for Three Related Condensed Cluster Halides of the Rare-Earth Elements. *Inorg. Chem.* **2010**, *49*, 9949–9957. [CrossRef] [PubMed]
74. Köckerling, M.; Martin, J.D. Electronic Fine Tuning of the Structures of Reduced Rare-Earth Metal Halides. *Inorg. Chem.* **2001**, *40*, 389–395. [CrossRef] [PubMed]
75. Bell, T. Seltenerd-Clusterkomplexe mit endohedralen Übergangsmetallatomen und ihre strukturellen sowie elektronischen Beziehungen zu intermetallischen Phasen. Dr. rer. nat. Thesis, University of Cologne, Dr. Hut Publishers, Munich, Germany, 2014.
76. Nesper, R. Bonding Patterns in Intermetallic Compounds. *Angew. Chem. Int. Ed. Engl.* **1991**, *30*, 789–817. [CrossRef]
77. Wade, K. The Structural Significance of the Number of Skeletal Bonding Electron-pairs in Carboranes, the Higher Boranes and Borane Anions, and Various Transition-metal Carbonyl Cluster Compounds. *J. Chem. Soc. D Chem. Commun.* **1971**, 792–793. [CrossRef]
78. Wade, K. Skeletal electron counting in cluster species. some generalisations and predictions. *Inorg. Nucl. Chem. Lett.* **1972**, *8*, 559–562. [CrossRef]
79. Lin, Q.; Corbett, J.D. Exploratory Syntheses and Structures of $SrAu_{4.3}In_{1.7}$ and $CaAg_{3.5}In_{1.9}$: Electron-Poor Intermetallics with Diversified Polyanionic Frameworks That Are Derived from the $CaAu_4In_2$ Approximant. *Inorg. Chem.* **2011**, *50*, 11091–11098. [CrossRef] [PubMed]
80. Pyykkö, P. Strong Closed-Shell Interactions in Inorganic Chemistry. *Chem. Rev.* **1997**, *97*, 597–636. [CrossRef] [PubMed]
81. Pyykkö, P.; Desclaux, J.-P. Relativity and the Periodic System of the Elements. *Acc. Chem. Res.* **1979**, *12*, 276–281. [CrossRef]
82. Pyykkö, P. Relativistic Effects in Structural Chemistry. *Chem. Rev.* **1988**, *88*, 563–594. [CrossRef]
83. Pyykkö, P. Theoretical Chemistry of Gold. *Angew. Chem. Int. Ed.* **2004**, *43*, 4412–4456. [CrossRef] [PubMed]
84. Smetana, V.; Lin, Q.; Pratt, D.K.; Kreyssig, A.; Ramazanoglu, M.; Corbett, J.D.; Goldman, A.I.; Miller, G.J. A Sodium-Containing Quasicrystal: Using Gold To Enhance Sodium´s Covalency in Intermetallic Compounds. *Angew. Chem. Int. Ed.* **2012**, *51*, 12699–12702. [CrossRef] [PubMed]
85. Gautier, R.; Zhang, X.; Hu, L.; Yu, L.; Lin, Y.; Sunde, T.O.L.; Chon, D.; Poeppelmeier, K.R.; Zunger, A. Prediction and accelerated laboratory discovery of previously unknown 18-electron ABX compounds. *Nat. Chem.* **2015**, *7*, 308–316. [CrossRef] [PubMed]

86. Steinberg, S.; Stoffel, R.P.; Dronskowski, R. Search for the Mysterious SiTe—An Examination of the Binary Si-Te System Using First-Principles-Based Methods. *Cryst. Growth Des.* **2016**, *16*, 6152–6155. [CrossRef]

87. Wang, F.; Pearson, K.N.; Miller, G.J. EuAg$_x$Al$_{11-x}$ with the BaCd$_{11}$-Type Structure: Phase Width, Coloring, and Electronic Structure. *Chem. Mater.* **2009**, *21*, 230–236. [CrossRef]

88. Steinberg, S.; Card, N.; Mudring, A.-V. From the Ternary Eu(Au/In)$_2$ and EuAu$_4$(Au/In)$_2$ with Remarkable Au/In Distributions to a New Structure Type: The Gold-Rich Eu$_5$Au$_{16}$(Au/In)$_6$ Structure. *Inorg. Chem.* **2015**, *54*, 8187–8196. [CrossRef] [PubMed]

89. Smetana, V.; Corbett, J.D.; Miller, G.J. Four Polyanionic Compounds in the K-Au-Ga System: A Case Study in Exploratory Synthesis and of the Art of Structural Analysis. *Inorg. Chem.* **2012**, *51*, 1695–1702. [CrossRef] [PubMed]

90. Smetana, V.; Miller, G.J.; Corbett, J.D. Three Alkali-Metal-Gold-Gallium Systems. Ternary Tunnel Structures and Some Problems with Poorly Ordered Cations. *Inorg. Chem.* **2012**, *51*, 7711–7721. [CrossRef] [PubMed]

91. Smetana, V.; Miller, G.J.; Corbett, J.D. Polycluster and Substitution Effects in the Na-Au-Ga System: Remarkable Sodium Bonding Characteristics in Polar Intermetallics. *Inorg. Chem.* **2013**, *52*, 12502–12510. [CrossRef] [PubMed]

92. Smetana, V.; Corbett, J.D.; Miller, G.J. Complex Polyanionic Nets in RbAu$_{4.01(2)}$Ga$_{8.64(5)}$ and CsAu$_5$Ga$_9$: The Role of Cations in the Formation of New Polar Intermetallics. *Z. Anorg. Allg. Chem.* **2014**, *640*, 790–796. [CrossRef]

93. Mishra, T.; Lin, Q.; Corbett, J.D. Gold Network Structures in Rhombohedral and Monoclinic Sr$_2$Au$_6$(Au,T)$_3$ (T = Zn, Ga). A Transition via Relaxation. *Inorg. Chem.* **2013**, *52*, 13623–13630. [CrossRef] [PubMed]

94. Palasyuk, A.; Grin, Y.; Miller, G.J. Turning Gold into "Diamond": A Family of Hexagonal Diamond-Type Au-Frameworks Interconnected by Triangular Clusters in the Sr-Al-Au System. *J. Am. Chem. Soc.* **2014**, *136*, 3108–3117. [CrossRef] [PubMed]

95. Li, B.; Kim, S.-J.; Miller, G.J.; Corbett, J.D. Synthesis, Structure, and Bonding in K$_{12}$Au$_{21}$Sn$_4$. A Polar Intermetallic Compound with Dense Au$_{20}$ and open AuSn$_4$ Layers. *Inorg. Chem.* **2009**, *48*, 11108–11113. [CrossRef] [PubMed]

96. Smetana, V.; Corbett, J.D.; Miller, G.J. Na$_8$Au$_{9.8(4)}$Ga$_{7.2}$ and Na$_{17}$Au$_{5.87(2)}$Ga$_{46.63}$: The diversity of pseudo 5-fold symmetries in the Na–Au–Ga system. *J. Solid State Chem.* **2013**, *207*, 21–28. [CrossRef]

97. Celania, C.; Smetana, V.; Provino, A.; Pecharsky, V.; Manfrinetti, P.; Mudring, A.-V. R$_3$Au$_9$Pn (R = Y, Gd-Tm; Pn = Sb, Bi): A Link between Cu$_{10}$Sn$_3$ and Gd$_{14}$Ag$_{51}$. *Inorg. Chem.* **2017**, *56*, 7247–7256. [CrossRef] [PubMed]

98. Lin, Q.; Corbett, J.D. Multiple Nonstoichiometric Phases with Discrete Composition Ranges in the CaAu$_5$-CaAu$_4$Bi-BiAu$_2$ System. A Case Study of the Chemistry of Spinodal Decomposition. *J. Am. Chem. Soc.* **2010**, *132*, 5662–5671. [CrossRef] [PubMed]

99. Smetana, V.; Steinberg, S.; Mudryk, Y.; Pecharsky, V.; Miller, G.J.; Mudring, A.-V. Cation-Poor Complex Metallic Alloys in Ba(Eu)-Au-Al(Ga) Systems: Identifying the Keys that Control Structural Arrangements and Atom Distributions at the Atomic Level. *Inorg. Chem.* **2015**, *54*, 10296–10308. [CrossRef] [PubMed]

100. Provino, A.; Steinberg, S.; Smetana, V.; Kulkarni, R.; Dhar, S.K.; Manfrinetti, P.; Mudring, A.-V. Gold-rich R$_3$Au$_7$Sn$_3$: Establishing the interdependence between electronic features and physical properties. *J. Mater. Chem. C* **2015**, *3*, 8311–8321. [CrossRef]

101. Lin, Q.; Vetter, J.; Corbett, J.D. Disorder-Order Structural Transformation in Electron-Poor Sr$_3$Au$_8$Sn$_3$ Driven by Chemical Bonding Optimization. *Inorg. Chem.* **2013**, *52*, 6603–6609. [CrossRef] [PubMed]

102. Lin, B.; Corbett, J.D. Different Cation Arrangements in Au-In Networks. Syntheses and Structures of Six Intermetallic Compounds in Alkali-Metal-Au-In Systems. *Inorg. Chem.* **2007**, *46*, 6022–6028. [CrossRef]

103. Lin, Q.; Mishra, T.; Corbett, J.D. Hexagonal-Diamond-like Gold Lattices, Ba and (Au,T)$_3$ Interstitials and Delocalized Bonding in a Family of Intermetallic Phases Ba$_2$Au$_6$(Au,T)$_3$ (T = Zn, Cd, Ga, In, or Sr). *J. Am. Chem. Soc.* **2013**, *135*, 11023–11031. [CrossRef] [PubMed]

104. Gerke, B.; Hoffmann, R.-D.; Pöttgen, R. Zn$_3$ and Ga$_3$ Triangles as Building Units in Sr$_2$Au$_6$Zn$_3$ and Sr$_2$Au$_6$Ga$_3$. *Z. Anorg. Allg. Chem.* **2013**, *639*, 2444–2449. [CrossRef]

105. Gerke, B.; Korthaus, A.; Niehaus, O.; Haarmann, F.; Pöttgen, R. Triangular Zn$_3$ and Ga$_3$ units in Sr$_2$Au$_6$Zn$_3$, Eu$_2$Au$_6$Zn$_3$, Sr$_2$Au$_6$Ga$_3$, and Eu$_2$Au$_6$Ga$_3$—Structure, magnetism, ^{151}Eu Mössbauer and $^{69;71}$Ga solid state NMR spectroscopy. *Z. Naturforsch. B* **2016**, *71*, 567–577. [CrossRef]

106. Gerke, B.; Pöttgen, R. Sr$_2$Au$_6$Al$_3$ and Eu$_2$Au$_6$Al$_3$—First Representatives of the Sr$_2$Au$_6$Zn$_3$ Type with Aluminium Triangles. *Z. Naturforsch. B* **2014**, *69*, 121–124. [CrossRef]

107. Smetana, V.; Steinberg, S.; Card, N.; Mudring, A.-V.; Miller, G.J. Crystal Structure and Bonding in BaAu$_5$Ga$_2$ and AeAu$_{4+x}$Ga$_{3-x}$ (Ae = Ba and Eu): Hexagonal Diamond-Type Au Frameworks and Remarkable Cation/Anion Partitioning in the Ae-Au-Ga Systems. *Inorg. Chem.* **2015**, *54*, 1010–1018. [CrossRef] [PubMed]

108. Dubois, J.-M.; Belin-Ferré, E. *Complex. Metallic Alloys: Fundamentals and Applications*; WILEY-VCH Publishers GmbH & Co. KGaA: Weinheim, Germany, 2011; ISBN 978-3-527-32523-8.

109. Urban, K.; Feuerbach, M. Structurally complex alloy phases. *J. Non. Cryst. Solids* **2004**, *334–335*, 143–150. [CrossRef]

110. Li, B.; Kim, S.-J.; Miller, G.J.; Corbett, J.D. Gold Tetrahedra as Building Blocks in K$_3$Au$_5$Tr (Tr = In, Tl) and Rb$_2$Au$_3$Tl and in Other Compounds: A Broad Group of Electron-Poor Intermetallic Phases. *Inorg. Chem.* **2009**, *48*, 6573–6583. [CrossRef] [PubMed]

111. Wuttig, M.; Yamada, N. Phase-change materials for rewriteable data storage. *Nat. Mater.* **2007**, *6*, 824–832. [CrossRef] [PubMed]

112. Lencer, D.; Salinga, M.; Wuttig, M. Design Rules for Phase-Change Materials in Data Storage Applications. *Adv. Mater.* **2011**, *23*, 2030–2058. [CrossRef] [PubMed]

113. Wuttig, M.; Raoux, S. The Science and Technology of Phase Change Materials. *Z. Anorg. Allg. Chem.* **2012**, *638*, 2455–2465. [CrossRef]

114. Bensch, W. Phasenwechselverbindungen auf Chalkogenidbasis für die optische und elektrische Datenspeicherung. *Z. Anorg. Allg. Chem.* **2008**, *634*, 2009. [CrossRef]

115. Chen, M.; Rubin, K.A.; Barton, R.W. Compound materials for reversible, phase-change optical data storage. *Appl. Phys. Lett.* **1986**, *49*, 502–504. [CrossRef]

116. Ohno, E.; Yamada, N.; Kurumizawa, T.; Kimura, K.; Takao, M. TeGeSnAu Alloys for Phase Change Type Optical Disk Memories. *Jpn. J. Appl. Phys.* **1989**, *28*, 1235–1240. [CrossRef]

117. Lencer, D.; Salinga, M.; Grabowski, B.; Hickel, T.; Neugebauer, J.; Wuttig, M. A map for phase-change materials. *Nat. Mater.* **2008**, *7*, 972–977. [CrossRef] [PubMed]

118. Shportko, K.; Kremers, S.; Woda, M.; Lencer, D.; Robertson, J.; Wuttig, M. Resonant bonding in crystalline phase-change materials. *Nat. Mater.* **2008**, *7*, 653–658. [CrossRef] [PubMed]

119. Esser, M.; Maintz, S.; Dronskowski, R. Automated first-principles mapping for phase-change materials. *J. Comput. Chem.* **2017**, *38*, 620–628. [CrossRef] [PubMed]

120. Deringer, V.L.; Zhang, W.; Lumeij, M.; Maintz, S.; Wuttig, M.; Mazzarello, R.; Dronskowski, R. Bonding Nature of Local Structural Motifs in Amorphous GeTe. *Angew. Chem. Int. Ed.* **2014**, *53*, 10817–10820. [CrossRef] [PubMed]

121. Wuttig, M.; Lüsebrink, D.; Wamwangi, D.; Welnic, W.; Gilleßen, M.; Dronskowski, R. The role of vacancies and local distortions in the desing of new phase-change materials. *Nat. Mater.* **2007**, *6*, 122–128. [CrossRef] [PubMed]

122. Matsunaga, T.; Yamada, N.; Kubota, Y. Structure of stable and metastable Ge$_2$Sb$_2$Te$_5$, an intermetallic compound in GeTe-Sb$_2$Te$_3$ pseudo-binary systems. *Acta Crystallogr. Sect. B* **2004**, *60*, 685–691. [CrossRef] [PubMed]

123. Chong, T.C.; Shi, L.P.; Zhao, R.; Tan, P.K.; Li, J.M.; Lee, H.K.; Miao, X.S.; Du, A.Y.; Tung, C.H. Phase change random access memory cell with superlattice-like structure. *Appl. Phys. Lett.* **2006**, *88*, 122114. [CrossRef]

124. Simpson, R.E.; Fons, P.; Kolobov, A.V.; Fukaya, T.; Krbal, M.; Yagi, T.; Tominaga, J. Interfacial phase-change memory. *Nat. Nanotechnol.* **2011**, *6*, 501–505. [CrossRef] [PubMed]

125. Buck, M.R.; Sines, I.T.; Schaak, R.E. Liquid-Phase Synthesis of Uniform Cube-Shaped GeTe Microcrystals. *Chem. Mater.* **2010**, *22*, 3236–3240. [CrossRef]

126. Konze, P.M.; Deringer, V.L.; Dronskowski, R. Understanding the Shape of GeTe Nanocrystals from First Principles. *Chem. Mater.* **2016**, *28*, 6682–6688. [CrossRef]

127. Tominaga, J.; Kolobov, A.V.; Fons, P.; Nakano, T.; Murakami, S. Ferroelectric Order Control of the Dirac-Semimetal Phase in GeTe-Sb2Te3 Superlattices. *Adv. Mater. Interfaces* **2014**, *1*, 1300027. [CrossRef]

128. Ohyanagi, T.; Kitamura, M.; Araidai, M.; Kato, S.; Takaura, N.; Shiraishi, K. GeTe sequences in superlattice phase change memories and their electrical characteristics. *Appl. Phys. Lett.* **2014**, *104*, 252106. [CrossRef]

129. Momand, J.; Wang, R.; Boschker, J.E.; Verheijen, M.A.; Calarco, R.; Kooi, B.J. Interface formation of two- and three-dimensionally bonded materials in the case of GeTe-Sb$_2$Te$_3$ superlattices. *Nanoscale* **2015**, *7*, 19136–19143. [CrossRef] [PubMed]

130. Stoner, E.C. Collective electron ferromagnetism. *Proc. R. Soc. Lond. A* **1938**, *165*, 372–414. [CrossRef]

131. Landrum, G.A.; Dronskowski, R. The Orbital Origins of Magnetism: From Atoms to Molecules to Ferromagnetic Alloys. *Angew. Chem. Int. Ed.* **2000**, *39*, 1560–1585. [CrossRef]

132. Dronskowski, R. Itinerant Ferromagnetism and Antiferromagnetism from the Perspective of Chemical Bonding. *J. Quantum. Chem.* **2004**, *96*, 89–94. [CrossRef]

133. Dronskowski, R.; Korczak, K.; Lueken, H.; Jung, W. Chemically Tuning between Ferromagnetnism and Antiferromagnetism by Combining Theory and Synthesis in Iron/Manganese Rhodium Borides. *Angew. Chem. Int. Ed.* **2002**, *41*, 2528–2532. [CrossRef]

134. Fokwa, B.P.T.; Lueken, H.; Dronskowski, R. Rational Synthetic Tuning between Itinerant Antiferromagnetism and Ferromagnetism in the Complex Boride Series $Sc_2FeRu_{5-n}Rh_nB_2$ ($0 \leq n \leq 5$). *Chem. Eur. J.* **2007**, *13*, 6040–6046. [CrossRef] [PubMed]

135. Brgoch, J.; Goerens, C.; Fokwa, B.P.T.; Miller, G.J. Scaffolding, Ladders, Chains, and Rare Ferrimagnetism in Intermetallic Borides: Electronic Structure Calculations and Magnetic Ordering. *J. Am. Chem. Soc.* **2011**, *133*, 6832–6840. [CrossRef] [PubMed]

136. Zhang, Y.; Miller, G.J.; Fokwa, B.P.T. Computational Design of Rare-Earth-Free Magnets with the $Ti_3Co_5B_2$-Type Structure. *Chem. Mater.* **2017**, *29*, 2535–2541. [CrossRef]

137. Fokwa, B.P.T.; Samolyuk, G.D.; Miller, G.J.; Dronskowski, R. Ladders of a Magnetically Active Element in the Structure of the Novel Complex Borid $Ti_9Fe_2Ru_{18}B_8$: Synthesis, Structure, Bonding, and Magnetism. *Inorg. Chem.* **2008**, *47*, 2113–2120. [CrossRef]

138. Eck, B.; Dronskowski, R.; Takahashi, M.; Kikkawa, S. Theoretical calculations on the structures, electronic and magnetic properties of binary 3d transition metal nitrides. *J. Mater. Chem.* **1999**, *9*, 1527–1537. [CrossRef]

139. Clark, W.P.; Steinberg, S.; Dronskowski, R.; McCammon, C.; Kupenko, I.; Bykov, M.; Dubronvinsky, L.; Akselrud, L.G.; Schwarz, U.; Niewa, R. High-Pressure NiAs-Type Modification of FeN. *Angew. Chem. Int. Ed.* **2017**, *56*, 7302–7306. [CrossRef] [PubMed]

MDPI

St. Alban-Anlage 66

4052 Basel

Switzerland

Tel. +41 61 683 77 34

Fax +41 61 302 89 18

www.mdpi.com

Crystals Editorial Office

E-mail: crystals@mdpi.com

www.mdpi.com/journal/crystals